SpringerBriefs in Climate Studies

SpringerBriefs in Climate Studies present concise summaries of cutting-edge research and practical applications. The series focuses on interdisciplinary aspects of Climate Science, including regional climate, climate monitoring and modeling, palaeoclimatology, as well as vulnerability, mitigation and adaptation to climate change. Featuring compact volumes of 50 to 125 pages (approx. 20,000- 70,000 words), the series covers a range of content from professional to academic such as: a timely reports of state-of-the art analytical techniques, literature reviews, in-depth case studies, bridges between new research results, snapshots of hot and/or emerging topics Author Benefits: SpringerBriefs in Climate Studies allow authors to present their ideas and readers to absorb them with minimal time investment. Books in this series will be published as part of Springer's eBook collection, with millions of users worldwide. In addition, Briefs will be available for individual print and electronic purchase. SpringerBriefs books are characterized by fast, global electronic dissemination and standard publishing contracts. Books in the program will benefit from easy-to-use manuscript preparation and formatting guidelines, and expedited production schedules. Both solicited and unsolicited manuscripts are considered for publication in this series. Projects will be submitted to editorial review by editorial advisory boards and/or publishing editors. For a proposal document please contact the Publisher.

More information about this series at https://link.springer.com/bookseries/11581

Tetsuji Ito • Makoto Tamura
Akihiko Kotera • Yuki Ishikawa-Ishiwata
Editors

Interlocal Adaptations to Climate Change in East and Southeast Asia

Sharing Lessons of Agriculture, Disaster Risk Reduction, and Resource Management

 Springer

Editors
Tetsuji Ito
College of Humanities and Social
Sciences/Global and Local Environment
Co-creation Institute
Ibaraki University
Ibaraki, Japan

Akihiko Kotera
Global and Local Environment
Co-creation Institute
Ibaraki University
Ibaraki, Japan

Makoto Tamura
Global and Local Environment
Co-creation Institute
Ibaraki University
Ibaraki, Japan

VNU Vietnam Japan University
Hanoi, Vietnam

Yuki Ishikawa-Ishiwata
Global and Local Environment
Co-creation Institute
Ibaraki University
Ibaraki, Japan

ISSN 2213-784X   ISSN 2213-7858   (electronic)
SpringerBriefs in Climate Studies
ISBN 978-3-030-81206-5   ISBN 978-3-030-81207-2   (eBook)
https://doi.org/10.1007/978-3-030-81207-2

This Springer imprint is published by the registered company Springer Nature Switzerland AG
The registered company address is: Gewerbestrasse 11, 6330 Cham, Switzerland

Foreword

We live in a rapidly changing world.

The significant factors causing environmental change today are the COVID-19 pandemic and climate change. We cannot yet foresee when COVID-19 will end, even though it has been spreading since its first report in December 2019. As of May 2021, more than 150 million people have been infected, and the number of COVID-related deaths has exceeded 3 million people worldwide. With such direct damages, COVID-19 has profoundly influenced the living conditions of vulnerable people such as those living in poverty, non-regular workers, the elderly, children, and women.

It has been pointed out that a potential cause of its occurrence and spread is the expansion of human activities, which degrades natural ecosystems and increases the risks of encountering unknown viruses. Recent globalization trends have contributed to the rapid diffusion of the virus and the expansion of the pandemic. The COVID-19 pandemic also highlights essential issues in today's society, such as the relationship between humans and nature, the vulnerable sides of globalization, and various socioeconomic divides among and within countries.

Climate change has also been progressing, making its impacts more apparent globally, particularly over the past 10 years. For example, India was hit by extraordinary heatwaves of more than 50°C in 2019, and the Philippines has suffered from severe damage from super typhoons. Simultaneously, Japan has been experiencing the increased impact of larger-scale floods caused by torrential rains and typhoons, which have caused escalating levels of damage. Bushfires in Australia in 2020 and California in 2021 gave a strong shock to the world. Although the times and places were different, all of these events were similarly reported using words such as "unprecedented" or "once a lifetime" occurrences. Hence, it is clear that among the features of today's rapidly changing world is the reality that environmental changes are confronting us with unparalleled challenges and risks.

In the face of this situation, world society has started to act. In terms of responding to climate change, 2015 was the epoch-making year when the Paris Agreement and Sustainable Development Goals (SDGs) were agreed upon in the international arena. The Paris Agreement set targets to hold the increase in the global average temperature to well below 2 °C above pre-industrial levels and pursue efforts to limit the temperature increase to 1.5 °C. Following this agreement, measures aimed at decarbonization, such as changing energy sources from fossil fuels to renewables and electric vehicle promotion, have advanced worldwide. In 2020, many countries declared to achieve carbon neutral by 2050.

These international trends have shown substantial advancements in response to the challenges of climate change. However, achieving the Paris goals will require time, which means that climate change and its effects will inevitably continue to increase for several decades. Therefore, we need another measure, i.e., adaptation, to increase preparedness for the adverse effects of climate change. To realize a safe, secure, climate-resilient, and sustainable society, the two pillars of mitigation (decarbonization) and adaptation are essential.

The shape and scale of climate change impacts vary with location. This is because climate change manifests itself differently from place to place, and because natural environments and societies are highly variable in each locality. As a result, each location requires specific adaptation approaches that correspond to their uniqueness. If we looked at the Asia and Pacific region, we see immense geographical, environmental, and socio-cultural variety. These include low-lying deltas vulnerable to sea-level rise, small islands with fragile ecosystems and traditional indigenous societies, high mountains and arid inland desert areas. They are all the homes of diverse communities and cultures. Therefore, it is necessary to craft suitable approaches to adaptation for each region.

This is why we need a participatory approach that includes a wide variety of local stakeholders for identifying problems, planning adaptation strategies, and implementing their options. This approach also makes it possible to meet other challenges facing local communities, such as agricultural and fisheries promotion, water resource management, while also addressing vitalization of local economy and socioeconomic divides. Since these efforts are all interconnected, they will require the adequate support of central governments and international institutes.

I sincerely expect this book *Interlocal Adaptations to Climate Change in East and Southeast Asia: Sharing Lessons of Agriculture, Disaster Risk Reduction, and Resource Management* to help us visualize the current impacts and levels of adaptation in the Asia and Pacific region, and I am sure it will contribute to sharing lessons acquired from them. Now that we have begun to advance climate change responses based on the two pillars of mitigation and adaptation, I hope this effort will clear and expand the way toward a sustainable future society.

Global and Local Environment   Nobuo Mimura
Co-creation Institute
Ibaraki University
Ibaraki, Japan

Preface

This book, titled *Interlocal Adaptations to Climate Change in East and Southeast Asia: Sharing Lessons of Agriculture, Disaster Risk Reduction, and Resource Management*, was written by researchers from Japan, Vietnam, Thailand, Indonesia, the Philippines, and China involved in climate change adaptation in East and Southeast Asia. Although each author has a different specialty, we have all agreed to engage in either interdisciplinary or even transdisciplinary collaboration to accomplish this project.

This book aims to achieve the following:

1. Promote interlocal lessons learned by sharing climate change adaptations, such as through "agriculture and natural resource management" and "disaster risk reduction and human resource development"
2. Develop new adaptation measures and research approaches that can consider the regional nature of East and Southeast Asia
3. Share practical adaptation options that have permeated society in each country/region

*   *   *

As is well known, Asia is one of the regions most vulnerable to the impacts of climate change, as more than 60% of the world's population lives in this region, making it the growth center of the world. Although Asia is composed of a wide variety of countries and incredibly vast territories, in this book, we focus on East and Southeast Asia, including Japan, based on our research network.

The term "climate crisis" offers a more comprehensive description of contemporary global conditions compared with "climate change," which might be insufficient. The current climate crisis not only alters the natural environment, but also affects the social environment and deeply impacts cultures and customs.

What should be done about the climate crisis? More importantly, what can be done? Although single individuals do not have substantial power, there are still many things that can be accomplished. We are researchers. We have the wisdom to make connections across countries and disciplines. Even if each individual possesses only marginal power, we believe that we can facilitate noticeable change by consolidating our connections and forming a network.

*   *   *

The term "interlocal," which is not very common in English, is one of the main keywords of this book. In relation to environmental issues, this term often refers to the concept of "Think globally, act locally." However, the emergence of environmental problems, in terms of both the natural and social environment, has not necessarily spread uniformly globally, and differs depending on the locality, such as countries or regions. Therefore, we believe it is necessary to "Think interlocally," which means to connect the dots (locality). At the same time, we believe it is also necessary to "Act interlocally." Researchers who have conducted research in local areas with different social practices can exchange knowledge with each other to connect these localities, transforming points into lines, lines into planes, and finally planes into three-dimensional objects. We envision such a composition. We believe that "sharing lessons" will allow us to obtain truly useful practical knowledge.

The focus of "adaptation" in this book does not mean that "mitigation" is not important. As is often said, "adaptation" and "mitigation" are similar to the wheels of a car. We believe that we can respond to climate change or the climate crisis by turning all wheels at the same time. However, it is clear based on reports from the Intergovernmental Panel on Climate Change (IPCC) that these problems cannot be solved by "mitigation" alone. Human beings have survived and built prosperity through "adaptation." Therefore, we now believe that the main question should be: "How can we 'adapt' to this situation?"

Regarding specific issues, this book focuses primarily on agriculture, disaster risk reduction, and resource management. Of course, these topics do not cover all related issues. We hope that sharing the lessons learned about these issues can serve as practical examples. As an additional note, "resources" include not only "natural resources," but also "human resources." We think that the term management for people (human resource management) is not always suitable, but we understand that it also refers to education, nurturing, and growth.

*   *   *

At Ibaraki University, to which we editors belong, an organization called the Institute for Global Change Adaptation Science (ICAS), which was established in 2006, has established a strong track record of research and education focusing on "climate change adaptation" from a relatively early stage. In recognition of this, Ibaraki University will be in charge of the master's program in climate change and development (MCCD) at VNU Vietnam Japan University (VJU), which opened in September 2018, and will be responsible for supervising the research conducted by many master's students in Vietnam. Based on the establishment of a base in Hanoi, the capital of Vietnam, our project has been adopted by the Core-to-Core Program of the Japan Society for the Promotion of Science (JSPS) with the theme of "Southeast Asia Research-based Network on Climate Change Adaptation Science" and promoted to expand networking in East and Southeast Asia. In April 2020, the organization of Ibaraki University evolved into the Global and Local Environment Co-creation Institute (GLEC).

*   *   *

In recent years, as the world started to come together to tackle such global issues, the emergence of ultra-nationalism has fostered the notion that global warming is a hoax, a narrative designed to divide people. We should not give credence to this idea, but combat it instead in solidarity.

The source of severe acute respiratory syndrome coronavirus 2 (SARS-CoV-2) that caused the coronavirus disease 2019 (COVID-19) pandemic in 2020 remains unclear according to a World Health Organization (WHO) survey, but climate change is a leading theory. Although scientific verification is needed to clarify this, the theory highlights the fact that climate change may impact numerous issues that are even more unexpected in the future.

Our network and the GLEC also want to carry out such work to "create a healthy environment together." May our small actions eventually build to become a big wave. Let's become planters of a "tree of hope" by collaborating together.

On behalf of the editors, Tetsuji Ito

Ibaraki, Japan

Tetsuji Ito
Makoto Tamura
Akihiko Kotera
Yuki Ishikawa-Ishiwata

Acknowledgement

This book was funded by Core-to-Core Program of the Japan Society for the Promotion of Science (JSPS) with the theme of "Southeast Asia Research-based Network on Climate Change Adaptation Science", Environment Research and Technology Development Fund (JPMEERF15S11413, JPMEERF20S11811) of the Environmental Restoration and Conservation Agency of Japan.

Contents

Contributors

Rizki Amalia Indonesian Oil Palm Research Institute, Kota Medan, North Sumatra, Indonesia

Orlando F. Balderama College of Engineering, Isabela State University, Isabela, Philippines

Dietriech Geoffrey Bengen Faculty of Fisheries and Marine Science, IPB University, Bogor, Indonesia

Rex Victor O. Cruz College of Forestry and Natural Resources, University of the Philippines Los Baños, College, Los Baños, Laguna, Philippines

Duong Thi Hong Nguyen Department of Climate Change and Sustainable Development, Hanoi University of Natural Resource and Environment, Hanoi, Vietnam

Jun Furuya Social Sciences Division, Japan International Research Center for Agricultural Sciences, Ibaraki, Japan

Josephine E. Garcia College of Forestry and Natural Resources, University of the Philippines Los Baños, College, Los Baños, Laguna, Philippines

Thu Hoai Nguyen VNU Vietnam Japan University, Hanoi, Vietnam

Hung The Nguyen Department of Climate Change and Sustainable Development, Hanoi University of Natural resource and Environment, Hanoi, Vietnam

Yuki Ishikawa-Ishiwata Global and Local Environment Co-creation Institute, Ibaraki University, Ibaraki, Japan

Tetsuji Ito College of Humanities and Social Sciences/Global and Local Environment Co-creation Institute, Ibaraki University, Ibaraki, Japan

Kazuyuki Kita Graduate School of Science and Engineering, Ibaraki University, Ibaraki, Japan

Akihiko Kotera Global and Local Environment Co-creation Institute, Ibaraki University, Ibaraki, Japan

Catherine C. De Luna Interdisciplinary Studies Center for Integrated Natural Resources and Environment Management, University of the Philippines Los Baños, College, Los Baños, Laguna, Philippines

Eri Matsuura College of Agriculture, Ibaraki University, Ibaraki, Japan

Yulu Ma College of Agriculture, Inner Mongolia Minzu University, Tongliao, China

Nobuo Mimura Global and Local Environment Co-creation Institute, Ibaraki University, Ibaraki, Japan

Satoshi Murakami Faculty of engineering, Fukuoka University, Fukuoka, Japan

Zulfi Prima Sani Nasution Indonesian Oil Palm Research Institute, Kota Medan, North Sumatra, Indonesia

Quang Van Nguyen VNU Vietnam Japan University, Hanoi, Vietnam

Ratnawati Nurkhoiry Indonesian Oil Palm Research Institute, Kota Medan, North Sumatra, Indonesia

Sachnaz Desta Oktarina Indonesian Oil Palm Research Institute, Kota Medan, North Sumatra, Indonesia

Tae Yoon Park Graduate School of Education, Yonsei University, Seoul, South Korea

Oanh Thi Pham VNU Vietnam Japan University, Hanoi, Vietnam

Florencia B. Pulhin College of Forestry and Natural Resources, University of the Philippines Los Baños, College, Los Baños, Laguna, Philippines

Juan M. Pulhin Interdisciplinary Studies Center for Integrated Natural Resources and Environment Management, University of the Philippines Los Baños, College, Los Baños, Laguna, Philippines
College of Forestry and Natural Resources, University of the Philippines Los Baños, College, Los Baños, Laguna, Philippines

Mark Anthony M. Ramirez Resources, Environment and Economics Center for Studies, Inc. (REECS), Quezon, Philippines

Nobuo Sakagami College of Agriculture, Ibaraki University, Ibaraki, Japan

Aiko Sakurai Department of Social Sciences, Toyo Eiwa University, Kanagawa, Japan
International Research Institute of Disaster Science, Tohoku University, Miyagi, Japan

Makoto Tamura Global and Local Environment Co-creation Institute, Ibaraki University, Ibaraki, Japan
VNU Vietnam Japan University, Hanoi, Vietnam

Maricel A. Tapia-Villamayor College of Forestry and Natural Resources, University of the Philippines Los Baños, College, Los Baños, Laguna, Philippines

Thi Thi My Tong Vietnam Institute of Economics, Vietnam Academy of Social Science, Hanoi, Vietnam

Tung Duy Do VNU Vietnam Japan University, Hanoi, Vietnam

Sukanya Vongtanaboon Faculty of Science and Technology, Phuket Rajabhat University, Phuket, Thailand

Kazuya Yasuhara Global and Local Environment Co-creation Institute, Ibaraki University, Ibaraki, Japan

Abbreviations

AHP	Analytical Hierarchy Process
AIWMP	Abuan Integrated Watershed Management Project
A-PLAT	Climate Change Adaptation Information Platform
AP-PLAT	Asia-Pacific Climate Change Adaptation Information Platform
ASTI	Advance Science and Technology Institute
BAPPENAS	Ministry of National Development Planning, Indonesia
BC	Black Carbon
BPTP	Balai Pengkajian Teknologi Pertanian
CBDRR	Community-Based Disaster Risk Reduction
CBMS	Community-Based Management System
CC	Climate Change
CCA	Climate Change Adaptation
CCS	Carbon Capture and Storage
CDP	Comprehensive Development Plan
CLUP	Comprehensive Land Use Plan
COP	Conference of Parties
CPI	Consumer Price Index
CRF	Climate Resilience Framework
CSA	Climate-Smart Agriculture
CSIRO	Commonwealth Scientific and Industrial Research Organization
CSTP	Committee for Scientific and Technological Policy
DENR	Department of Environment and Natural Resources
DILG	Department of Local Government
DOST	Department of Science and Technology
DOST-ASTI	Advanced Science and Technology Institute of the DOST
DPL	Community-Based Marine Protected Area
DREAM	Disaster Risk and Exposure Assessment for Mitigation Project
DRR	Disaster Risk Reduction
DRRM	Disaster Risk Reduction and Management
DRRMO	Disaster Risk Reduction and Management Office
DRRTHG	Disaster Risk Reduction X Treasure Hunting Game

DSSAT	Decision Support System for Agro-Technology Transfer
EWS	Early Warning Systems
FAO	Food and Agricultural Organization
FAO-STAT	Food and Agricultural Organization Corporate Statistical Databases
FDSS	Farmer Decision Support System
GAP	Good Agricultural Practices
GCM	Global Circulation Model
GEJE	2011 Great East Japan Earthquake
GHG	Green House Gas
GSO	General Statistical Office of Vietnam
HCM	Ho Chi Minh City
HEC-HMS	Hydrologic Engineering Center-Hydrologic Modeling System
HEC-RAS	Hydrologic Engineering Center-River Analysis System
IBM	International Business Machines
IBM-WEDA	Weather Data Solutions of IBM
ICT	Information and Communication Technology
iLCCAC	Ibaraki Local Climate Change Adaptation Center
InSAR	Interferometry Synthetic Aperture Radar
IPCC	Intergovernmental Panel on Climate Change
IRR	Implementing Rules and Regulation
ISET	Institute for Social and Environmental Transition-International
IUCN	International Union for Conservation of Nature
JICA	Japan International Cooperation Agency
KCDA	Katahira Community Development Association
LCCAC	Local Climate Change Adaptation Center
LCCAP	Local Climate Change Action Plans
LDCs	Least Developed Countries
LDRRMC	Local Disaster Risk and Reduction Management Council
LGC	Local Government Code
LGU	Local Government Unit
LIDAR	Light Detection and Raging
LS	Land Subsidence
M&E	Monitoring and Evaluation
MCCD	Master's Program in Climate Change and Development
MCD	Centre for Marinelife Conservation and Community Development
MEXT	Ministry of Education, Culture, Sports, Science and Technology in Japan
MFAJ	Ministry of Foreign Affairs in Japan
MLIT	Ministry of Land, Infrastructure, Transportation and Tourism in Japan
MOA	Memorandum of Agreement
MOEJ	Ministry of the Environment in Japan
MONRE	Ministry of Natural Resources and Environment
MOU	Memorandum of Understanding
MRD	Mekong River Delta

NAPA	National Adaptation Programmes of Action
NCCAP	The Philippine's National Climate Change Action Plan
NDRRMC	National Disaster Risk Reduction and Management Council
NOAH	Nationwide Operational Assessment of Hazards Project
NPAs	National Adaptation Plans
PHP	Philippines Peso
$PM_{2.5}$	Particulate Matter with diameter smaller than 2.5μm
RA	Republic Act
RAN-API	Rencana Aksi National Adaptasi Perubahan Iklim
RCP	Representative Concentration Pathways
R-DRRMP	Reconstruction and DRR Mapping Program
SDGs	Sustainable Development Goals
SFDRR	Sendai Framework for Disaster Risk Reduction
SLCPs	Short-Lived Climate Pollutants
SLR	Sea Level Rise
SME	Small-Medium Enterprise
SMS	Short Messaging System
SNS	Social Networking Service
SRES	Special Report on Emission Scenarios
SRI	System of Rice Intensification
SSP	Shared Socioeconomic Pathways
Sub-NIAPP	Sub-National Institute of Agricultural Planning and Projection
SRTM3	Shuttle Radar Topography Mission, Ver.3
TO_3	Tropospheric ozone
UAVs	Unmanned Aerial Vehicles
UNDESA	United Nations Department of Economic and Social Affairs
UNFCCC	United Nations Framework Convention on Climate Change
UNISDR	United Nations International Strategy for Disaster Reduction
US EPA	United States Environmental Protection Agency
USDA PS&D	United States Department of Agriculture, Production, Supply and Distribution
WB	World Bank
WDI	World Development Indicator
WISE	Weather Information-Integration for System Enhancement
WLR	Water Level Rise
XTNP	Xuan Thuy National Park

Chapter 1
Climate Change Risk and Adaptation

Makoto Tamura

1.1 The Impacts of Climate Change in East and Southeast Asia

Climate change is considered by many to be the most critical issue of our time, posing a threat to security and socio-economic prosperity at the global level (MFAJ 2017). Asia is very vulnerable to the impacts of climate change, as more than 60% (approximately 4.5 billion) of the world's people live in the region, making it a growth center of the world (UNDESA 2017).

In 2017, human-induced global warming reached approximately 1 °C (between 0.8 °C and 1.2 °C) above pre-industrial levels, increasing at a rate of 0.2 °C (likely between 0.1 °C and 0.3 °C) per decade (Fig. 1.1; IPCC 2018). Figure 1.2 shows projected temperature changes in Southeast Asia, as calculated by Japan's Ministry of the Environment and the Japan Meteorological Agency (MOEJ 2015). It shows expected temperature differences between the recent past (1984–2004) and future climate conditions (2080–2100) in the region. The colors indicate the average values for the RCP2.6, RCP4.5, and RCP6.0 scenarios, and nine RCP8.5 scenarios. In all scenarios, temperatures show a distinct increasing tendency. These increasing temperatures will have serious physical and socio-economic impacts.

Climate change poses a variety of threats. Figure 1.3 shows a map of the risks attributable to climate change and attendant socio-economic circumstances. The former includes increasing rainfall, storms, flooding, inundation, sea level rise (SLR), uncertainty in terms of agricultural production, and the occurrence of heat waves. To illustrate such risks, Fig. 1.4 shows the potentially inundated areas of

M. Tamura (✉)
Global and Local Environment Co-creation Institute, Ibaraki University, Ibaraki, Japan

VNU Vietnam Japan University, Hanoi, Vietnam
e-mail: makoto.tamura.rks@vc.ibaraki.ac.jp

© The Author(s) 2022
T. Ito et al. (eds.), *Interlocal Adaptations to Climate Change in East and Southeast Asia*, SpringerBriefs in Climate Studies,
https://doi.org/10.1007/978-3-030-81207-2_1

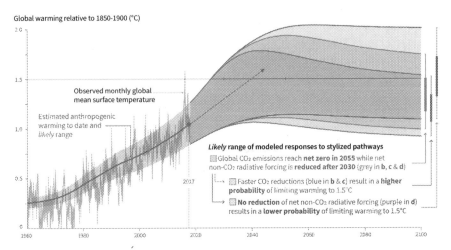

Global warming relative to 1850-1900 (°C)

Fig. 1.1 Observed global temperature change and modeled responses to stylized anthropogenic emission and forcing pathways (IPCC 2018)

Asia due to SLR in 2100 under RCP8.5/MIROC-ESM (Tamura et al. 2019). Lowland areas, such as the Mekong Delta, the Yangtze River Delta, and the Ganges Delta may be partially or heavily inundated by SLR. Indeed, China, Canada, Vietnam, the United States, Brazil, Australia, Indonesia, and India all face similar challenges in terms of having the world's largest potentially inundated areas. Included among the associated socio-economic issues are urbanization, population growth, increased migration, income disparity, volatile food prices, lack of insurance schemes, lack of financial resources to prepare for extreme events, and impact of information technology.

1.2   Climate Change Responses: Mitigation and Adaptation

There are two approaches to addressing the issue of environmental change: one is to remove the causes of the change; the other is to adopt measures that will allow societies to adjust to the adverse effects. In the context of climate change, these responses are referred to as mitigation and adaptation, respectively. Mitigation to reduce greenhouse gas (GHG) emissions and their role in climate change include energy conservation and the development of alternative energy sources, as well as forest protection and afforestation management. Adaptation, which serves to adjust human and natural systems to the assumed ongoing climate change, might include measures such as disaster prevention, as well as making changes in the cultivation of plant species and breeding new plant varieties.

Mitigation strategies are roughly divided into two categories: those that reduce the sources of GHG emissions and those that act as GHG sinks. Reducing GHG emissions would include improved energy efficiency in both supply and demand, as

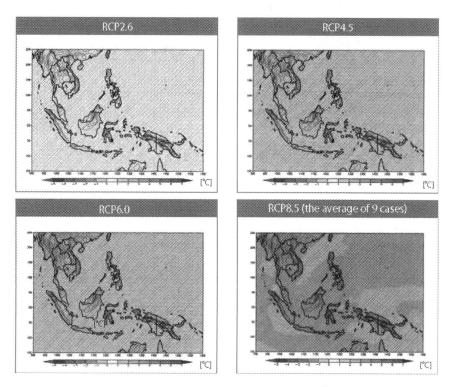

Fig. 1.2 Changes in annual mean surface temperatures (°C) in Southeast Asia under each RCP scenario (MOEJ 2015)

well as the use of technologies for reducing GHG emissions. Supply-related reduction measures would include the development and widespread use of alternative energy derived from non-fossil fuels. Demand-related measures would involve energy conservation at various stages, including the production stage, the transportation stage, and the domestic utilization stage. Enhancing GHG sinks includes increasing GHG absorption by ecosystems through afforestation, re-forestation, forest management, and carbon capture and storage (CCS) or sequestration. Measures such as afforestation and appropriate forest management clearly contribute positively to the conservation of ecosystems; however, CCS is somewhat problematic because of its potential to adversely impact ecosystems.

Adaptation measures include the following (Hay and Mimura 2006):

- Avoiding or reducing the likelihood of adverse events or conditions. This means taking preventive measures against anticipated effects, e.g., improving catchment management, and avoiding excessive runoff and flooding.
- Reducing consequences. This involves measures to diminish damage that has already occurred, e.g., ensuring healthy reef and mangrove systems, which act as buffers during storm surges.

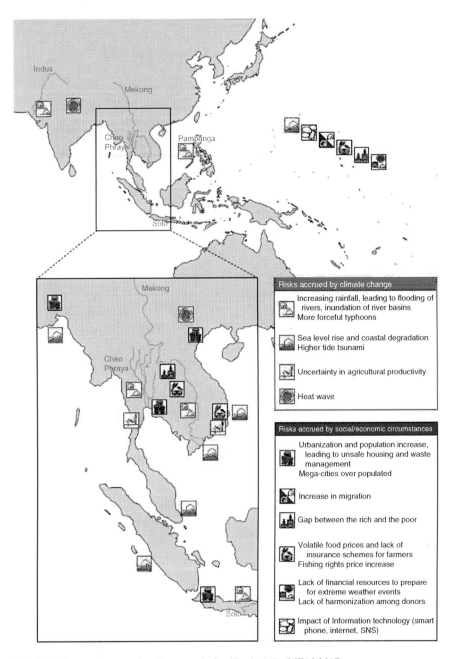

Fig. 1.3 Climate change and socioeconomic fragility in Asia (MFAJ 2017)

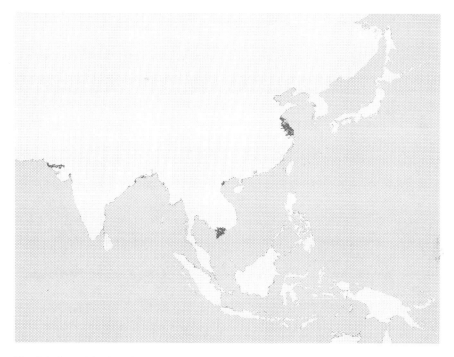

Fig. 1.4 Potentially inundated areas in Asia due to sea level rise in 2100 under the RCP8.5/ MIROC-ESM. (Adopted from Tamura et al. 2019)

- Re-distributing or sharing risks. This includes measures to lessen the costs of damage by dispersing them among many people or over a long period, e.g., insurance schemes.
- Accepting risk. This means doing nothing, at least for a particular time, but includes the opportunity to learn from the experience.

It generally takes considerable time for mitigation measures to take effect, but they can provide wide-ranging benefits. In contrast, adaptation measures have a rather immediate effect, but tend to operate in limited areas. Although mitigation measures can be evaluated on the basis of GHG emissions, it is difficult to set similar baseline and result indicators for adaptation measures and to properly evaluate their effectiveness. Both approaches have specific advantages and can be viewed as complementary.

1.3   Vulnerability, Sensitivity, and Resilience

Figure 1.5 illustrates temporal changes in vulnerability corresponding to mitigation of, and adaptation to, climate change. Mitigation is intended to control the climate's external forces (hazards) while adaptation is intended to increase resilience or

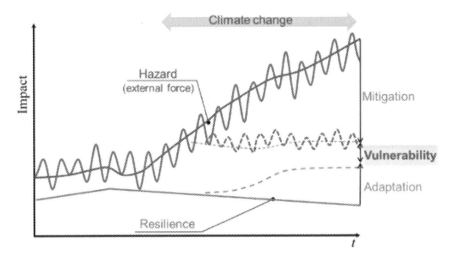

Fig. 1.5 Temporal change in vulnerability corresponding to mitigation and adaptation to climate change (Komatsu et al. 2013)

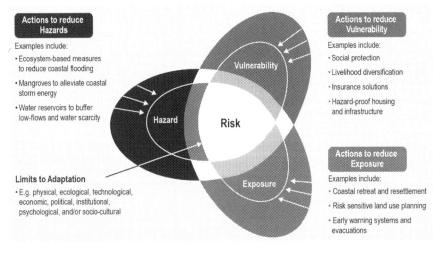

Fig. 1.6 Interaction among exposure, vulnerability, and their adaptations (IPCC 2019)

adaptive capacity. The risk of climate-related impacts is a product of the interaction of climate-related hazards and the vulnerability and exposure of humans and natural systems (Fig. 1.6). Here, hazards refer to threats, and includes both hazardous events and trends. Exposure refers to the presence of people and assets in places that could be adversely affected.

Vulnerability to climate change is determined by (1) the magnitude of external forces, such as an increase in air temperature, SLR, and changing rainfall patterns; (2) the susceptibility of nature and society to these external forces; and (3) the capacity of society to adapt to these external forces. Accordingly, societies that are easily damaged by climate change are considered to be highly vulnerable. Thus, the combination of larger climate hazards (external forces) and less adaptive capacity (less resilience) means higher vulnerability. Even under conditions where there is the same level of hazard, the risk will be dependent on the local situation. If the region has small exposure (e.g., slightly affected population and small assets) and good adaptive capacity to climate change, then the risk will not be so serious (and vice versa).

1.4 Approaches and Categories of Adaptation

1.4.1 Approaches

Two main approaches to adaptation have been developed to address adaptation: a top-down scientific approach and a bottom-up regional approach (Fig. 1.7). The scientific or top-down approach involves long-term adaptation measures by both national and local governments, and includes climate projections, their downscaling, impact/vulnerability assessments, and the planning of adaptation (e.g., Klein et al. 1999). In contrast, regional approach or community-based adaptation addresses challenges at the local level and seeks to promote the participation of stakeholders (especially those in the local community) in the process of formulating adaptive measures. This regional or bottom-up approach encourages people in the community to recognize their future risks and to participate in the planning and

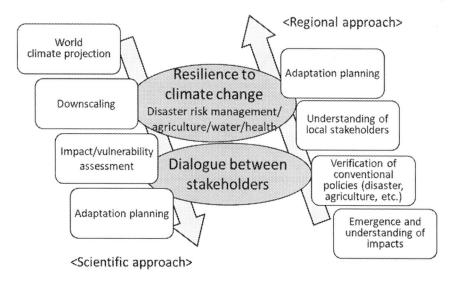

Fig. 1.7 Dual approach for climate change adaptation. (Revised from Tamura et al. 2014)

implementation of adaptation measures (e.g., Adger et al. 2005). Dialogue between stakeholders may bridge these two approaches.

This book covers both research and activities among some East and Southeast Asia countries so that it can fill in the gap between two approaches including agriculture, disaster risk management, resource management, and human resource development.

1.4.2   Categories

Based on the timescale (short-term, mid-term, and long-term), adaptation measures can be further categorized as follows: (1) efforts to strengthen existing measures related to adaptation, (2) adaptive management for middle and long-term impacts, and (3) measures aimed at a fundamental improvements in sensitivity (Fig. 1.8). Enhancing adaptive capacity is the main objective at the initial level of adaptation, while improvement in sensitivity, including reconstruction and transformation, is the main goal at the latter level.

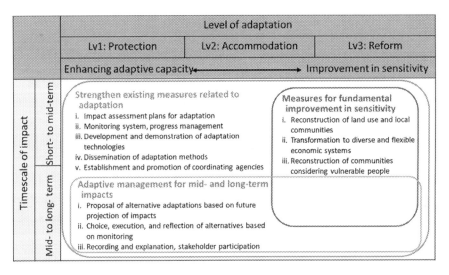

Fig. 1.8 Level and time scale of adaptation (Shirai et al. 2014)

1.5 Trends in International Policy and National Adaptation Plans (NAPs)

The Paris Agreement, adopted in December 2015 at the 21st Conference of the Parties (COP21) by the UN Framework Convention on Climate Change (UNFCCC), has a shared international long-term goal of "holding the increase in the global average temperature to well below 2°C above pre-industrial levels" and sets the direction toward net zero anthropogenic GHG emissions by the second half of the twenty-first century. With respect to adaptation, the agreement calls for "enhancing adaptive capacity, strengthening resilience and reducing vulnerability to climate change", and urges the parties to "engage in a process to formulate and implement national adaptation plans."

The UNFCCC began its National Adaptation Programmes of Action (NAPA) in 2001 to support the efforts of the Least Developed Countries (LDCs) in addressing their urgent and immediate need to cope with the impacts of climate change. In 2010, the UNFCCC established the NAP process at COP16 in Cancun. Comparing the two approaches, NAPA is rather urgent and reactive and focuses on actions for LDCs to be taken primarily based on observed or past events. The NAP process is more proactive and, in addition to observed information, relies on future climate projections and their likely impacts (UNFCCC 2012). Of 153 developing countries, 91 have initiated the NAP process and 11 have submitted NAPs as of 2018 (UNFCCC 2018). Since the adverse impacts of climate change have been recognized by nearly all countries, NAPs are not limited to LDCs, but are promoted by both developing and developed countries, including the EU countries (UK, Germany, etc.), the USA, Japan, South Korea, and China. Japan formulated its "National Plan for Adaptation to the Impacts of Climate Change" in November 2015 (Cabinet decision 2015) and enacted the "Climate Change Adaptation Act" in June 2018. After formulating their NAP, national governments need to develop methodologies for monitoring and evaluating (M & E) the progress of their adaptation efforts.

1.6 Key Elements of a National-Level Adaptation Plan/ Strategy

National-level adaptation plays a key role in adaptation planning and implementation. It serves to coordinate adaptation responses at subnational and local levels, where diverse processes and outcomes are called for (IPCC 2014). National-level coordination includes the provision of information about potential risks in order to strengthen the actions of state and local governments. These activities provide policy frameworks that guide decisions at the subnational level, coordinating the creation of legal frameworks, directing sectoral action and targeting resources for national development (agriculture, fisheries, health, ecosystem protection, among others), protecting vulnerable groups, and providing financial support for the various levels of government.

As adaptation activities have progressed, multiple challenges have emerged, including how to manage the decision-making process, how to develop effective strategies and plans, and how to implement them. In this regard, individual roles within a multilevel governance system have become an issue made more complex by the need for horizontal coordination among different agencies and departments, and vertical coordination among various stakeholders, from national to regional to local actors (IPCC 2014). National governments need to coordinate and enhance appropriate multilevel adaptations. Accordingly, they should (1) share common understanding among the line ministries in order to ensure that concerted actions are taken with minimum effort and cost, and that duplication is avoided, (2) prioritize actions according to evidence, recognizing the limitations of time and budget, and (3) allocate a portion of the national budget to adaptation policy, that is sufficient to assess the potential effectiveness of adaptation measures.

UNFCCC (2012) summarizes the steps associated with each of the elements of national adaptation plan formulation for LDCs (Table 1.1). One of several key milestones under Element A is the establishment of institutional arrangements that undertake coordination and leadership in the process, including internal coordination with line ministries and role allotment. Element B includes key activities related

Table 1.1 Steps under each of the elements of the formulation of national adaptation plans (UNFCCC 2012)

Element A. lay the groundwork and address gaps
1. Initiating and launching of the NAP process
2. Stocktaking: Identifying available information on climate change impacts, vulnerability and adaptation and assessing gaps and needs of the enabling environment for the NAP process
3. Addressing capacity gaps and weaknesses in undertaking the NAP process
4. Comprehensively and iteratively assessing development needs and climate vulnerabilities

Element B. preparatory elements
1. Analyzing current climate and future climate change scenarios
2. Assessing climate vulnerabilities and identifying adaptation options at the sector, subnational, national and other appropriate levels
3. Reviewing and appraising adaptation options
4. Compiling and communicating national adaptation plans
5. Integrating climate change adaptation into national and subnational development and sectoral planning

Element C. implementation strategies
1. Prioritizing climate change adaptation in national planning
2. Developing a (long-term) national adaptation implementation strategy
3. Enhancing capacity for planning and implementation of adaptation
4. Promoting coordination and synergy at the regional level and with other multilateral environmental agreements

Element D. reporting monitoring and review
1. Monitoring the NAP process
2. Reviewing the NAP process to assess progress, effectiveness and gaps
3. Iteratively updating the national adaptation plans
4. Outreach on the NAP process and reporting on progress and effectiveness

to technical assessments that enable a country to make informed decisions based on scientific methodologies and findings. Impact assessment, prioritized impacts, adaptation measures, internal coordination with line ministries, and role allotment are commonly required. The activities under Element C of the process comprise the implementation strategies for NAPs. Element D, reporting, monitoring and review- ing the process of formulating and implementing NAPs, entails collecting informa- tion on the process, assessing it through a national monitoring and evaluation (M&E) system and providing outputs for reporting on progress to the COP. The guidelines suggest that the National Adaptation Plan process should be designed to be flexible and non-prescriptive; hence, countries may apply the suggested steps based on their particular circumstances, choosing those steps that add value to their planning process and sequencing activities. Figure 1.9 shows the progress made in the development of the NAP process from 2015 to 2018.

Adaptation planning and implementation are dynamic iterative learning pro- cesses that recognize the complementary roles of adaptation strategies, plans, and actions at different levels (national, subnational, and local) (IPCC 2014). Iterative risk management is a useful framework for decision-making in complex situations characterized by large potential consequences, persistent uncertainties, long time- frames, potential for learning, and multiple climatic and non-climatic influences changing over time (Fig. 1.10). People and knowledge shape the process and its outcomes.

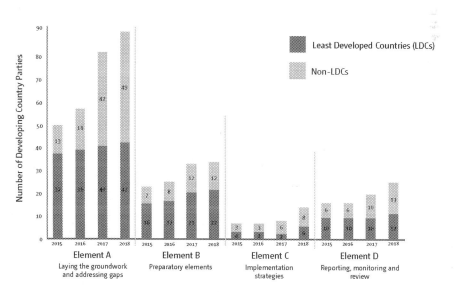

Fig. 1.9 Progress made by developing countries in the process used to formulate and implement national adaptation plans by process element from 2015–2018 (FCCC/SBI/2018/INF.13) (UNFCCC 2018)

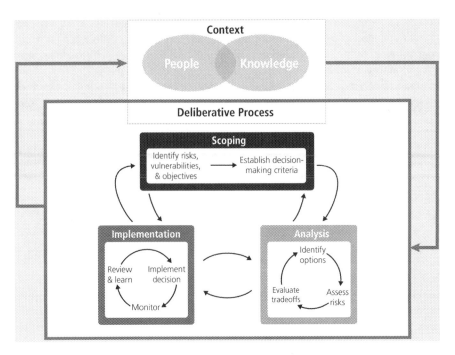

Fig. 1.10 Climate-change adaptation as an iterative risk management process with multiple feedbacks (IPCC 2014)

1.7 Adaptation in Local Governments: Japan's Case

NAPs can serve as the main driver to promote regional adaptation or guide the actions taken by local governments. The impacts of climate change, vulnerabilities, and priorities vary substantially depending on regional characteristics. Each local government should consider adaptation measures proactively and address them carefully.

The Japanese government and research communities have conducted extensive studies on the projected impacts of climate change and have investigated suitable adaptation measures (MOEJ 2008, 2010; MEXT et al. 2009; CSTP 2010). The Ministry of Environment (MOEJ 2008) has introduced the principle of "wise adaptation" based on the following concepts:

1. As a policy development operating under uncertainty, adaptation should be based on an effective, efficient and flexible approach. In spite of significant progress in research and policies, uncertainties still exist in the projections of future climate change, the associated impacts and social trends. Because rapid advances are occurring in global observations and climate projections, adaptation plans that can be revised every few years should be adopted, as opposed to implementing unchangeable measures.

2. Wise adaptation considers climate change adaptation in the wider context of sustainability and the well-being of society. Adaptation to climate change should contribute to other social goals, such as mitigation of climate change, the creation of an environmentally friendly, safe and secure society in accordance with sustainable development goals (SDGs).

The Japanese Diet enacted Japan's "Climate Change Adaptation Act" in June 2018, according to which the national government shall formulate an NAP to promote adaptation in all sectors. Methodologies for monitoring and evaluating (M & E) the progress of adaptation efforts need to be developed.

Climate change impacts and vulnerabilities depend on local conditions. The implementation of adaptive measures is mainly conducted by local communities consistent with the overall planning of national adaptations. As shown in Fig. 1.11, the climate change adaptation act of Japan recommends that prefectures and municipalities assign "local climate change adaptation center (LCCAC)" as local climate change data collection and provision centers. As of August 2021, 34 among 47 prefectures and 9 cities have established local centers. For example, Ibaraki Prefecture established Ibaraki Local Climate Change Adaptation Center (iLCCAC) at Ibaraki University in April 2019 (https://www.ilccac.ibaraki.ac.jp/). A-PLAT (http://www.adaptation-platform.nies.go.jp), which is organized by national climate change adaptation center, is providing national data and compiling local data. LCCAC provides more locally specific information and support for adaptation planning in local municipalities. These horizontal and vertical collaboration efforts are expected to

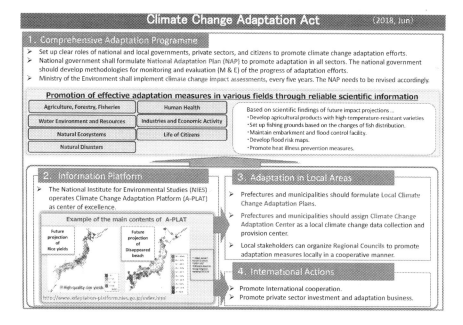

Fig. 1.11 Climate change adaptation act in Japan (MOEJ 2018)

fill the gap between scientific and regional approaches for adaptation, as shown in Fig. 1.7, and are significant for promoting resilient adaptations.

1.8 Lessons Learned

Developing countries with high vulnerability that face serious adverse impacts from climate change will need to increase their adaptive capacities. Long-term targets would include the ability to apply their own traditional experiences and knowledge of natural disasters, and to develop the capacity to implement their own monitoring and prediction techniques. To enhance these adaptive capacities, it is critical to establish a social system that can promote information collection and to share and raise awareness of the importance of these activities. Similar methodologies have been implemented to promote regional development. Indeed, adaptation is sometimes regarded as a means to accomplish development objectives, while development can be seen as the means to achieve adaptation objectives (McGray et al. 2007). In other words, because of the overlap between adaptation and regional development, increasing adaptive capacity depends, to a significant extent, on regional development paths. In this sense, it is difficult to separate actions to reduce vulnerability to climate change from those promoting sustainable development when there are limited resources that can be allocated to both action streams. Not surprisingly, some of the SDGs adopted at the Sustainable Development Summit in 2015 recognize this close connection. In particular, the targets for SDG 13 "Climate action (Take urgent action to combat climate change and its impacts)" include strengthening resilience and adaptive capacity to deal with climate hazards (United Nations 2018). Thus, decision makers have come to realize that 'mainstreaming' or incorporating adaptation policies into existing socioeconomic policies without distinguishing between climate change policies and, for example, land use and agriculture planning, is a logical evolution of the preferred solution.

Adaptation and mitigation serve as twin pillars in the battle against climate change. Together, they represent the main strategies for improving social resilience to climate change, ensuring human security and promoting sustainable development. As part of the battle plan, it is essential to review and reconstruct our national policies on land use, environmental use and city/rural planning. The ideal of a low-carbon and resilient society needs to be aggressively promoted so that a sustainable society can ultimately be created. The high-level target is to adopt a flexible response to climate change in order to sustain a dynamic and vital society. Importantly, wise adaptation to climate change must be implemented according to regional realities, changing regional and social modalities while simultaneously producing solutions to other pressing problems.

References

Adger WN, Hughes TP, Folke C, Carpenter SR, Rockstrom J (2005) Social-ecological resilience to costal disasters. Science 309(5737):1036–1039

Cabinet Decision in Japan (2015) National Plan for adaptation to the impacts of climate change. Cabinet Decision on 27 November 2015

CSTP (Council for Science and Technology Promotion) (2010) Planning technological development towards realizing a society adapting to climate change (Final Report). Task Force for Planning Technological Development towards Realizing a Society Adapting to Climate Change, The Cabinet Office, 28p (in Japanese)

Hay J, Mimura N (2006) Supporting climate change vulnerability and adaptation assessments in the Asia-Pacific region: an example of sustainability science. Sustain Sci 1(1):23–35

IPCC (2014) Climate change 2014 -impacts, adaptation and vulnerability: working group II contribution to the fifth assessment report of the IPCC. Cambridge University Press, London

IPCC (2018) Special report on global warming of 1.5 °C

IPCC (2019) Special report on the ocean and cryosphere in a changing climate

Klein RJT, Nicholls RJ, Mimura N (1999) Coastal adaptation to climate change: can the IPCC technical guidelines be applied? Mitig Adapt Strateg Glob Chang 4(3–4):239–252

Komatsu T, Shirai N, Tanaka M, Harasawa H, Tamura M, Yasuhara K (2013) Adaptation philosophy and strategy against climate change-induced geo-disasters. In: Proceedings of 10th JGS. Symposium on Environmental Geotechnics, 76–82

McGray H, Hammill A, Bradley R (2007) Weathering the storm: options for framing adaptation and development. World Resource Institute, Washington DC, 57p

MEXT (Ministry of Education, Culture, Sports, Science and Technology), JMA (Japan Meteorological Agency), and MOE (Ministry of Environment) (2009) Synthesis report on observations, projections, and impact assessments of climate change climate change and its impacts in Japan. MEXT, 74p (in Japanese)

MFAJ (2017) Analysis and proposal of foreign policies regarding the impact of climate change on fragility in the Asia-Pacific region- with focus on natural disasters in the region. Ministry of Foreign Affairs in Japan, 49p

MOEJ (2008) Wise adaptation to climate change, the committee on climate change impacts and adaptation research. Ministry of Environment in Japan, 70p

MOEJ (2010) Approaches to climate change adaptation. The committee on approaches to climate change adaptation, Ministry of Environment in Japan, 80p

MOEJ (2015) Climate change in Southeast Asia: outputs from GCM. Ministry of Environment in Japan, 12p

MOEJ (2018) Climate change adaptation act. https://www.env.go.jp/en/earth/cc/adaptation.html

Shirai N, Tanaka M, Tamura M, Yasuhara K, Harasawa H, Komatsu T (2014) Building and verification of a theoretical framework for climate change adaptation: strategies of climate change adaptation. Environ Sci 27(5):313–323. (in Japanese)

Tamura M, Yasuhara K, Shirai N, Tanaka M (2014) Wise adaptation to climate change: Japan's case. In: Prutsch A, McCallum S, Grothmann T, Swart R, Chauser I (eds) Climate change adaptation manual: lessons learned from European and other industrialized countries. Routledge, pp 314–319

Tamura M, Yotsukuri M, Kumano N, Yokoki H (2019) Global assessment of the effectiveness of adaptation in coastal areas based on RCP/SSP scenarios. Clim Chang 152(3–4):363–377

UNDESA (2017) World population prospects: the 2017 revision. United Nations, New York

UNFCCC (2012) National adaptation plans: technical guidelines for the National Adaptation Plan Process. LDC expert group, 148p

UNFCCC (2018) National Adaptation Plans 2018: Progress in the process to formulate and implement National Adaptation Plans. LDC expert group, 33p

United Nations (2018) The sustainable development goals report 2018, 40p

Part I
Agriculture and Natural Resource Management

Chapter 2
Participatory Climate Change Adaptation Using Watershed Approach: Processes and Lessons from the Philippines

Juan M. Pulhin, Maricel A. Tapia-Villamayor, Josephine E. Garcia, Catherine C. De Luna, Rex Victor O. Cruz, Florencia B. Pulhin, and Mark Anthony M. Ramirez

2.1 Introduction

The Philippine's National Climate Change Action Plan (NCCAP) aims primarily to build the adaptive capacity of local communities and increase the resilience of natural ecosystems to climate change to promote climate-risk resilience. Climate change is expected to have far-reaching impacts on the structure, ecological conditions, and functions of various ecosystems, diminishing their capacity to support people and communities. Thus, it is important to anticipate the conditions of a socio-ecological system considering climate change, and to maintain its integrity through adaptation.

The watershed approach, which uses the watershed as a planning unit and exemplifies a coordinated environmental management that focuses on public and private

J. M. Pulhin (✉)
Interdisciplinary Studies Center for Integrated Natural Resources and Environment Management, University of the Philippines Los Baños, College, Los Baños, Laguna, Philippines

College of Forestry and Natural Resources, University of the Philippines Los Baños, College, Los Baños, Laguna, Philippines
e-mail: jmpulhin@up.edu.ph

M. A. Tapia-Villamayor · J. E. Garcia · R. V. O. Cruz · F. B. Pulhin
College of Forestry and Natural Resources, University of the Philippines Los Baños, College, Los Baños, Laguna, Philippines

C. C. De Luna
Interdisciplinary Studies Center for Integrated Natural Resources and Environment Management, University of the Philippines Los Baños, College, Los Baños, Laguna, Philippines

M. A. M. Ramirez
Resources, Environment and Economics Center for Studies, Inc. (REECS), Quezon, Philippines

© The Author(s) 2022
T. Ito et al. (eds.), *Interlocal Adaptations to Climate Change in East and Southeast Asia*, SpringerBriefs in Climate Studies,
https://doi.org/10.1007/978-3-030-81207-2_2

sector efforts (US EPA 2008), ensures a holistic method of managing ecosystems. Designing strategies to sustainably use and manage the resources in the watershed in view of changes in climate constitute an important step towards moderating the potential impacts of climate change and taking advantage of the opportunities that climate change presents. This also contributes towards building the resilience of the people and various ecosystems in the watershed.

In the Philippines, watershed areas are under the administration of the Department of Environment and Natural Resources (DENR) and management is under the control of respective local government units (LGU). Nevertheless, watershed management entails collective action or a public involvement to influence collective action. This means different stakeholders, particularly in communities, mobilizing themselves for this common cause, or a public entity acting as a champion to pursue a common goal.

This paper presents the processes and lessons learned in developing a participatory climate change framework in two watersheds in the Philippines, namely Baroro Watershed and Saug Watershed. In the end, the project came up with a protocol on participatory climate change adaptation using watershed approach to enhance resilience of communities and ecosystems.

The Baroro Watershed is located in the northeastern part of the province of La Union and encompasses the municipalities of San Gabriel, San Juan, Bagulin, Bacnotan and Santol, and the city of San Fernando. It comprises a total of 19,486 hectares. The watershed is the main source of water for both irrigation and domestic purposes in all of the municipalities and city, except for Bagulin and Santol. Agriculture is the main source of income in the watershed.

The Saug Watershed, on the other hand, is located in the provinces of Davao del Norte and Compostela Valley. The watershed comprises the municipalities of Asuncion, Kapalong, New Corella and San Isidro and the city of Tagum in Davao del Norte, and the municipalities of Laak, Mawab, Monkayo, Montevista and Nabunturan in Compostela Valley. It has a total land area of 99,866 hectares, 60% of which is occupied by Davao del Norte and 40% by Compostela Valley. The watershed is an economically important catchment that hosts the agricultural production in the two provinces and supplies these areas with water, particularly the low-lying municipalities. Figure 2.1 shows the location of the two watersheds.

2.2   Planning for Resilience: An Integrated Approach

The complexity of the climate change problem demands an approach that would tackle its dynamic, multi-sectoral, multi-scalar and highly variable impacts. The risks and uncertainties associated with it also takes planning beyond the traditional 'predict and act' framework (Institute for Social and Environmental Transition-International (ISET) 2013), and strives for a system that is prepared for any disturbance or sudden change. Hence, national strategic plans for climate change and

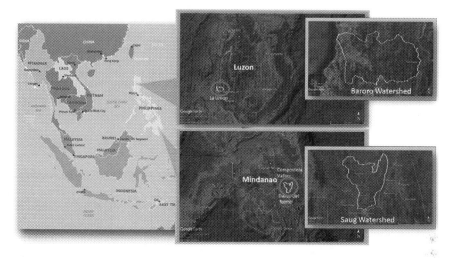

Fig. 2.1 Locations of Baroro and Saug watersheds

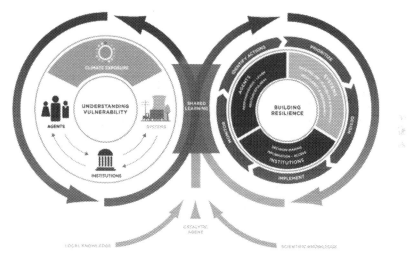

Fig. 2.2 ISET climate resilience framework (2014)

disasters envision resilience as the ultimate goal of all actions and efforts in response to these phenomena.

In 2013, ISET designed a Climate Resilience Framework (CRF) as part of measures intended to build resilience (Fig. 2.2). The framework emphasized that building resilience needs to occur in the context of a socio-ecological system so as to ensure that the dynamic interactions of different system elements and their feedbacks are considered. The CRF is a collaborative planning process based on understanding the vulnerability of the core components of a system – agents, institutions, and systems (left loop in Fig. 2.2) and strategic planning for climate change that highlights the resilient characteristics of each system component (right loop in Fig. 2.2), guided by an iterative learning approach that considers both local and scientific knowledge.

Capturing the concepts and principles of the ISET framework for building resilience, and considering also that managing climate and disaster-related risks benefit from an integrated approach rather than "separately managing individual types of risk or risk in particular locations" (IPCC 2012), a watershed approach was selected as the project framework for enhancing the resilience of communities and ecosystems in these project sites.

Using the watershed as a framework for planning not only provides a holistic approach that recognizes the linked social and ecological systems in a geographic unit, as well as its multifunctionality. This approach also benefits from "sound management techniques based on strong science and data" (US EPA 2008). From a mere hydrologic boundary from whence the concept was coined, it evolved to incorporate human use and economic values giving rise to Integrated Watershed Resource Management, which then expanded into a governance framework as encapsulated in the "watershed approach" (Cohen and Davidson 2011). US EPA (2008) defined watershed approach as an "environmental management that focuses public and private sector efforts to address the highest priority problems within hydrologically-defined geographic areas, taking into consideration both ground and surface water flow".

The use of the watershed approach also benefits other aspects of watershed management, such as streamflow regulation, conservation of soil resources, enhancement of soil infiltration capacity, soil erosion minimization, optimum production of goods and services, eradication of upland poverty, and environmental stabilization (Cruz 1999), that contribute to improving the integrity of various ecosystems and strengthen their natural resilience to climate change and disasters.

2.3 Adaptation as a Process

Adaptation is defined as the process of adjustment to actual or expected climate and its effects. In human systems, it is the process of adjustment to actual or expected climate stimuli and its effects in order to moderate harm or exploit beneficial benefits. In natural systems, it is the process of adjustment to actual climate and its effects; human intervention may facilitate adjustment to expected climate (IPCC 2014a).

Adaptation is implemented at various scales and levels, and its implementation differs depending on the context, such as resources, values and needs. Hence, the focus on adaptation as a process is important. However, current approaches to climate change adaptation prioritizes its technological and sectoral aspects, and little discussion is available on how it is practiced, as well as the impacts of adaptation strategies implemented (O'Brien and Hochachka 2010; IPCC 2014b).

Treating adaptation as a process, the project documented the steps it undertook to develop a protocol for participatory climate change adaptation. Following the principle that "all responses to climate change rely on information about risk and vulnerability" (IPCC 2012), the project conducted biophysical, socioeconomic, institutional, vulnerability and risk assessments that aimed to determine the current situation in the Baroro and Saug watersheds. The assessments relied on participatory rural appraisal techniques: spatial (fragmentation) analysis, water and greenhouse gas modeling; collection of secondary data for the watersheds' biophysical characteristics; household surveys; and in-depth interviews of municipal/city officers.

Consistent with the watershed approach, communities living across the different gradients of the watershed (upstream, midstream and downstream) were selected as samples, particularly for the participatory rural appraisal techniques and household surveys. A key strategy of the project was to harvest the local knowledge of the communities and municipal officers through a historical situational analysis of the watershed. This narrative focused on the changes in specific ecosystem services: freshwater production; soil productivity; food, fiber and raw materials; maintenance and biodiversity; cultural services; and micro-climate. The process, together with other participatory techniques, served as an eye-opener to the communities on the challenges currently faced by the watershed, its impacts on them, and the acknowledgement that climate change could bring greater danger to their already fragile ecosystems.

Through several workshops and seminars, the results of the above activities were presented to the communities and the different watershed stakeholders. These venues served as sites for integration of local and scientific knowledge, as the project team also explained the biophysical and socioeconomic assessments based on the data gathered and computer modeling, and the leaders (mayors) in the municipalities and cities within the watershed shared their current programs and projects which concern the watershed.

Based on these assessments, both Baroro and Saug watersheds were found to be approaching critical ecological limit based on the significant degrees of fragmentation from 1988 to 2015, notably from agriculture and urban expansion. Both watersheds were also found to be highly at risk to flooding, which is aggravated by siltation of the river systems, and the Saug Watershed is also facing serious erosion problems. Institutions were also found to have limited knowledge on climate change and even the use of the watershed as a planning unit, hence the lack of coordination with other municipalities in implementing environment and disaster-related strategies. Such lack of coordination has the potential to result in conflicts among the municipal/city leaders.

With the knowledge of the current situation of the watersheds, a visioning exercise was performed among the various stakeholders to ascertain the desired situation that they would like to achieve for the watershed in the future. Strategies that would also lead to the attainment of the desired future were also identified in a workshop environment. These stakeholders crafted their visions and encapsulated their strategies in a 'brand name' that represents the values and opinions of the communities and stakeholders for the watershed.

After working with the community and the municipal government in the watersheds, the project team sought an audience with the members of the Provincial Councils in each of the watersheds. All of the findings on the status and condition of the watershed, including the ways forward that were desired by the communities and municipalities to attain the vision they crafted, were also presented. This activity, particularly in the case of Baroro Watershed, persuaded the provincial government to unite and mend the differences among the municipal/city leaders and act as one for the benefit of the integrity of the watershed and the local communities that depend on it. A Memorandum of Agreement (MOA) and Memorandum of Understanding (MOU) were also signed by the leaders in both Baroro and Saug watersheds, respectively, to institutionalize their commitment and support for the action plans and to promote resilience in the watershed in the face of environmental degradation and climate change. A seed fund was even earmarked for these activities in the Baroro Watershed by the Provincial Council.

The legal frameworks, as represented by the MOA and MOU, provided the basis for the different stakeholders at the community and municipal levels, including other relevant agencies, to craft an action plan for climate change adaptation. This details the strategies, potential sources of funds, responsible groups or stakeholders, and the timeline for implementation. A monitoring and evaluation system was developed to ensure that the objectives of the plans are satisfied, leading to the attainment of the vision for the watershed. Pilot communities were also identified to implement the plan, and it is expected that this would ripple to the other communities as their impacts become visible, finally leading to the scaling up of the approach.

2.4   Demystifying Participation

Developing an adaptation plan to be led and piloted by the communities is not an easy feat. Nevertheless, giving them the capacity for such task is an indispensable aspect of adaptation as communities are at the forefront of impacts, and therefore, of the actions to mitigate climate change.

Community participation, similar to adaptation, is also a process. It refers to the process in which individuals and communities engage in decisions about things that affect their lives (Burns et al. 2004). Community-based adaptation to climate change is characterized by the following (Dodman and Soltesova 2012):

- Based on the premise that local communities have skills, experience, knowledge and networks to undertake locally appropriate activities to increase resilience;
- Recognizes limits to/failure of planned, 'top down' approaches to adaptation;

- Generates adaptation strategies through participatory processes involving local stakeholders—recognizes the need to include vulnerable people in decisions; and
- Builds on existing cultural norms and addresses local development issues that underlie vulnerability.

Guided by the above principles, the project explored what it takes to have successful community participation in environmental or development projects. In a workshop environment, stakeholders, including community members, were asked to share projects that exemplified successful participatory approaches. Concepts that emerged from their narratives were highlighted and synthesized to form guidelines for participation that the project could implement.

In the Baroro Watershed, the concepts related to participation that emerged from the narratives included: different actors; participation of enablers (i.e., the LGU as represented by the mayor) and the influencers (selected community members who will first join the project); identification of different organizations that could provide support; knowledge enhancement that leads to the realization and acceptance of the problem; and integration of livelihoods. All of these were summed up into four principles: (1) setting up a framework to guide participatory action; (2) sustainable action through enabling policies and institutional arrangement; (3) capacity building and communication planning; and (4) innovative financing. Similar concepts emerged during the discussion with communities and stakeholders in the Saug Watershed, but with an emphasis on incentives and livelihoods.

The framework for participatory action represents who would champion the cause and the availability of willing community members who could demonstrate how the project operates, as well as the benefits that could be obtained from it. This framework emphasizes that community-based adaptation is not just a private act, but more often than not, a public, influence-based collective action. Enabling policies and institutional arrangements guarantee the permanence of the initiative, as well as the support across different levels of governance (i.e., provincial, city/municipal, *barangay*/community). This is where the dialogues with the Provincial Council and their seal of commitment as represented by MOA/MOU were instrumental. Hence, both bottom-up and top-down approaches were necessary to lend legitimacy and unanimity for a community-based project.

Integral to adaptation planning is the change in knowledge and values of the communities through awareness raising and capacity building. This enhances the adaptive capacity of the communities through better understanding of these phenomena affecting them. Their involvement in each level of the participation process also empowered them and made their voices heard in planning. Lastly, innovative financing, which should also consider the livelihood of residents and other incentives, represent the primary tangible benefits that makes participation more worthwhile for the communities.

All of the above should be encapsulated in a "brand" that the communities can rally behind, that represents the values that they can relate to, and the vision that they aspire to for the watershed. In the Baroro Watershed, they labeled the participatory climate change adaptation as "Ipon Ti Baroro Watershed". This brand represents a fish that is endemic to the Baroro River with the same name, and which is an ecological indicator of the river ecosystem. Further, "ipon" also stands for unity ("pagtitipon" in Tagalog), and also connotes saving for the future. Meanwhile, the "branding" coined for Saug Watershed was the acronym SAUG, which stands for "sustainable approaches for unlimited goods and godly services".

2.5   A Protocol for Participatory Climate Change Adaptation Using Watershed Approach

Based on the above activities, a protocol was developed that summarized the steps and achievements for enhancing participation in climate change adaptation using a watershed approach. It depicts, through the title, the framework used in the process (i.e., the watershed approach), the adaptation planning steps (Steps 1–7), and the cross-cutting principles on participation and capacity enhancement followed in its implementation (Fig. 2.3).

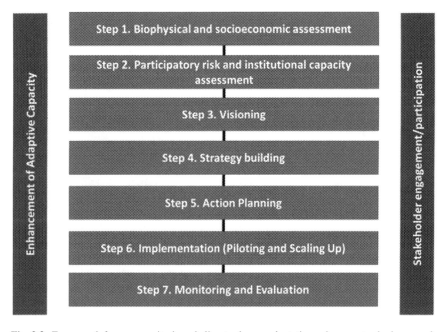

Fig. 2.3 Framework for community-based climate change adaptation using a watershed approach

2.5.1 Biophysical and Socio-economic Assessment

This step was guided by a systems approach, i.e., the watershed approach, that was used to analyze the different biophysical (climate, land use, soil and geology, water, biodiversity) and socio-economic (population, livelihood, policies and institutions) components of the watershed, and how these affect ecosystem services in the watershed. The process relied on several assessment methodologies, such as, carbon stocks assessment, water modeling, fragmentation analysis, household surveys, historical situational analysis and key informant interviews. The design of methodology for this step is not limited to the above examples, but would depend on the goals of the adaptation project, particularly the ecosystem functions and services that require strengthening.

2.5.2 Participatory Risk and Institutional Capacity Assessment

Participatory risk assessment involved the identification of hazards and their probability of occurrence, impacts, and the effectiveness of current adaptation strategies. It is an investigative method that employs a variety of participatory tools to engage local stakeholders in their own climate change-risk and vulnerability diagnosis. The institutional capacity assessment, on the other hand, guided by the elements of resilient institutions in the ISET framework, determined the access rights and entitlements, information flows, application of new knowledge, decision-making processes, as well as the capacity to anticipate risk, capacity to respond, and capacity to recover and change, of the different institutions in the watershed.

Included under this step is the participatory scenario development, which explored the future with the communities and other stakeholders in a creative and policy-relevant way. Given the current risks and impacts of climate-related events, the results of climate, water, and carbon stock modeling were presented, and this provided planners with a glimpse of what they should consider when developing planning frameworks.

2.5.3 Visioning

A vision represents the future that the stakeholders would like to achieve for themselves and the watershed. In a workshop environment, participants were asked to discuss how they wanted to see the watershed in the future and to write these visions on metacards (one item per metacard). All metacards were collected and similar responses were grouped. The top responses were placed together and used to create a vision statement, which was further refined by the participants until they reached a consensus.

The vision statements developed for the Baroro and Saug watersheds are as follows:

- Baroro Watershed – We envision the Baroro Watershed to be a sustainable and climate-resilient source of water, capable of providing a sufficient livelihood, and an ecotourism destination in La Union.
- Saug Watershed – A well-managed, climate-resilient Saug watershed sustaining multiple uses of goods and services for its constituents through science-based approaches and local participation.

2.5.4 Strategy Building

Strategy building refers to the actual strategies formulated by different stakeholders to respond to issues and problems identified in the assessment, considering future climate change and progress towards achieving the vision that they aspire to achieve. Strategies were identified per sector or ecosystem, and, considering the watershed gradient (upstream, mid-stream and downstream), were consolidated for consistency.

2.5.5 Action Planning

Action planning, following the legal framework emanating from the provincial government, entailed the detailed community-based adaptation planning for the different pilot communities. At this stage, commitment was already secured from different stakeholders, particularly the communities, to effectively implement the proposed action plan. Potential sources of funds are also identified. In the case of Baroro Watershed, part of the MOA stipulated seed funding for this initiative.

Some important strategies identified for the pilot communities were reforestation using native species, organic farming, regulation of harmful agricultural inputs, fish pen regulation, alternative livelihood creation, and solid waste management. This action plan is encapsulated in a "brand" that represents the overall values and objectives of the initiative.

2.5.6 Implementation (Piloting and Scaling Up)

Implementation included setting up of a pilot *barangay* or community for each watershed site to implement the participatory climate change adaptation. It requires the recognition and assistance of the LGU to ensure that the *barangay* has a legal identity, availability of resources, and to secure guarantees for the continuity. An organization structure and specific work plan were laid out, including the Monitoring and Evaluation System, to guide the implementation of the adaptation project. This part, however, was not covered by the project. Nevertheless, mechanisms were installed to regularly assess the status/progress of the community-based adaptation, as well as plans for potential scaling up.

2.5.7 Monitoring and Evaluation

Indicators for monitoring and evaluating the participatory community-based adaptation were developed from parameters measured before the project. This would then be compared to the future status of the watershed following implementation of the project. The indicators included: stream discharge, meteorological (rainfall and temperature), soil moisture, water use efficiency, biodiversity and ecosystems, water balance, and socioeconomic characteristics. Scale-appropriate indicators for the communities were also developed, and this would also be implemented using participatory approaches.

2.6 Lessons Learned

The project succeeded in formulating a protocol for participatory community-based adaptation using a watershed approach. This protocol contained the critical know-how in climate change adaptation and watershed management, particularly in improving the integrity of the ecosystems and ensuring the continuity of the goods and services that they provide for the welfare of the people.

The following are the lessons learned in the process of formulating the protocol:

- Continuing ecosystems degradation increases communities' risks and vulnerability to climate-induced hazards and disasters and undermines their resiliency.
- Recognizing the different scales of adaptation through the watershed approach (communities, municipal, and provincial levels) should consider each group's role —i.e., the context of their adaptation (values, resources, and needs).
- A participatory framework, such as that described above, is important for catalyzing collective action among stakeholders to enhance climate change adaptation.
- Solutions-based analysis that incorporates local knowledge will empower the community to be more mindful, prepared and proactive in addressing the potential negative impacts of climate change acting on their person, their livelihood and the environment.
- Recognizing the roles of communities in watershed management positively affects their status and stimulates their creativity in crafting adaptation strategies that are effective and bring about rehabilitation.
- Local stakeholders need recognition and assistance from LGUs, national agencies and other organizations to enable them to perform as effective watershed stewards.

- Local communities are willing to participate in implementing adaptation strategies that will conserve watersheds as well as their livelihoods.
- Lastly, community-based adaptation in the context of participatory watershed management cannot operate solely at the local level; it needs to be effectively linked to the higher scales of governance in order to enhance the resilience of communities and ecosystems.

References

Burns D, Heywood F, Taylor M, Wilde P, Wilson M (2004) Making community participation meaningful: a handbook for development and assessment. The Policy Press, Bristol. Retrieved from URL: https://www.jrf.org.uk/sites/default/files/jrf/migrated/files/jr163-community-participation-development.pdf

Cohen A, Davidson S (2011) The watershed approach: challenges, antecedents, and the transition from technical too to governance unit. Water Altern 4(1): 1–14. Retrieved from URL: http://www.water-alternatives.org/index.php/allabs/123-a4-1-1/file

Cruz RVO (1999) Integrated land use planning and sustainable watershed management. J Philippine Dev, Number 47 26(1) First Semester 1999. Retrieved from URL: https://dirp3.pids.gov.ph/ris/pjd/pidsjpd99-11and.pdf

Dodman D, Soltesova K (2012) Community-based adaptation in urban areas: potential and limits. Paper presented during the Third Global Forum on Urban Resilience and Adaptation. Session D1: Mitigating and adapting from the bottom up: community-based solutions. Retrieved from URL: https://www.slurc.org/uploads/1/0/9/7/109761391/dodman_-_cba_potential_and_limits.pdf

Institute for Social and Environmental Transition (ISET) – International (2013) Climate resilience framework: putting resilience into practice. ISET-International, Boulder, 22 p

Intergovernmental Panel on Climate Change (IPCC) (2012) Managing the risks of extreme events and disasters to advance climate change adaptation. In: Field CB, Barros V, Stocker TF, Qin D, Dokken DJ, Ebi KL, Mastrandrea MD, Mach KJ, Plattner G-K, Allen SK, Tignor M, Midgley PM (eds) A special report of working groups I and II of the intergovernmental panel on climate change. Cambridge University Press, Cambridge, UK/New York, NY, p 582

Intergovernmental Panel on Climate Change (IPCC) (2014a) Annex II: Glossary. Climate change 2014. Mach KJ, Planton S, von Stechow C (eds) Synthesis Report. Contribution of Working Groups I, II and III to the Fifth Assessment Report of the Intergovernmental Panel on Climate Change. [Core Writing Team, RK Pachauri and LA Meyer (eds)]. IPCC, Geneva, Switzerland, pp 117–130

Intergovernmental Panel on Climate Change (IPCC) (2014b) Climate change 2014: impacts, adaptation and vulnerability–summary for policymakers. Contribution of Working Group II to the Fifth Assessment Report of the Intergovernmental Panel on Climate Change: Field CB, Barros VR, Dokken DJ, Mach KJ, Mastrandrea MD, Billir TE, Chatterjee M, Ebi KL, Estrada YO, Genva RC, Girma B, Kissel ES, Levy AN, MacCracket S, Mastrandrea PR, White LL (eds) World Meteorological Organization, Geneva, Switzerland, 34. pp

O'Brien K, Hochachka G (2010) Integral adaptation to climate change. J Integral Theory Pract 5(1):89–102

PROVIA (2013) PROVIA guidance on assessing vulnerability, impacts and adaptation to climate change. Consultation Document, United Nations Environment Programme, Nairobi, 198 p

United States Environmental Protection Agency (USEPA) (2008) Watershed approach framework. https://www.epa.gov/sites/production/files/2015-06/documents/watershed-approach-framework.pdf

Chapter 3
Climate Change Adaptation Practices Towards Sustainable Watershed Management: The Case of Abuan Watershed in Ilagan City, Philippines

Orlando F. Balderama

Abbreviations

AIWMP	Abuan Integrated Watershed Management Project
ASTI	Advance Science and Technology Institute
CBMS	Community-Based Management System
LGU	Local Government Unit
CCA	Climate Change Adaptation
CDP	Comprehensive Development Plan
CLUP	Comprehensive Land Use Plan
DILG	Department of Local Government
DREAM	Disaster Risk and Exposure Assessment for Mitigation Project
DRRM	Disaster Risk Reduction and Management
DRRMO	Disaster Risk Reduction and Management Office
DSSAT	Decision Support System for Agro-Technology Transfer
FDSS	Farmer Decision Support System
IBM	International Business Machines
ICT	Information and Communication Technology
IRR	Implementing Rules and Regulation
LCCAP	Local Climate Change Action Plans
LDRRMC	Local Disaster Risk and Reduction Management Council
LGC	Local Government Code
NOAH	Nationwide Operational Assessment of Hazards Project
RA	Republic Act
SMS	Short Messaging System

O. F. Balderama (✉)
College of Engineering, Isabela State University, Isabela, Philippines
e-mail: orlando.f.balderama@isu.edu.ph

© The Author(s) 2022
T. Ito et al. (eds.), *Interlocal Adaptations to Climate Change in East and Southeast Asia*, SpringerBriefs in Climate Studies,
https://doi.org/10.1007/978-3-030-81207-2_3

3.1 Introduction

In the Philippines, it is the local government units (LGUs) who are at the forefront in implementing initiatives related to disaster risk reduction and management (DRRM) and climate change adaptation (CCA) in their respective jurisdictions. The legal framework governing DRRM and CCA devolves to the LGUs the responsibility to prepare and integrate local CCA and DRRM into locally mandated plans, particularly into the Comprehensive Land Use Plan (CLUP) and Comprehensive Development Plans (CDP). The legal framework governing the country's climate change and disaster risk reduction and management policies are as follows:

(a) The *Local Government Code (LGC) of 1991 (Republic Act [RA] No. 7160)*. The LGC devolves to LGUs the responsibility of delivering basic services in agriculture, health, environment, and social services to the local constituents. The Act mandates each LGU to prepare their CDP and CLUP.

(b) The *Climate Change Act of 2010 (RA 9729)* and its *Implementing Rules and Regulations (IRR) (Administrative Order 2010–01)*. The Act acknowledges the Philippines' vulnerability to climate change and the need for appropriate adaptation. This Act creates a comprehensive framework for systematically integrating climate change and disaster risk reduction (DRR) into various phases of policy formulation, development plans, poverty reduction strategies, and other development tools and techniques. The Act establishes the Climate Change Commission as the sole policymaking body to prepare a National Climate Change Framework, National Climate Change Action Plan, and guidelines for the preparation of local climate change action plans (LCCAP).

(c) *Disaster Risk Reduction and Management Act of 2010 (RA 10121) and IRR*. Section 11 (b)(2) of the Act mandates the Local Disaster Risk and Reduction Management Council (LDRRMC) to ensure that DRR and CCA are integrated into local development plans, programs, and budgets as a strategy in sustainable development and poverty reduction. This provision integrates DRRM and CCA into physical and land-use planning, budget, infrastructure, education, health, environment, housing, and other sectors. The law requires LGUs to set aside 5% of their regular revenues for DRRM.

This chapter discusses highlights and lessons learned from the Abuan Integrated Watershed Management Project (AIWMP), a five-year program funded by the United States Agency for International Development-Philippines, which focused on the implementation of bottom-up, self-initiated CCA strategies in the watershed. The initiated measures have led to policy outcomes that have enhanced the resiliency of communities and watershed ecosystems against natural disasters and climate change. In addition, the activities, institutional arrangements, and policy outcomes are also discussed. Innovations in the use of information communication technology (ICT), mobile technologies, and remote sensing offer opportunities to modernize the agriculture sector; however, barriers to upscaling require the establishment of an enabling policy environment and capacity building in order to extend the potential application of these technologies.

3.2 Project Setting: Climate Outlook, Ilagan City and the Abuan Watershed

Isabela province in the Cagayan Valley is the Philippines' top corn producer. Ilagan City in the province of Isabela has the biggest land area under corn and is regarded as the Corn Capital of the Philippines. Climate change poses long-term threats to the livelihoods of farmers and to national food security. According to the Philippine Atmospheric Geophysical and Astronomical Services Administration (2011), the country's weather bureau, the 2050 scenario in Isabela will result in a 1.9 °C–2.1 °C increase in temperature, a 29% decrease in mean rainfall in the dry months, and a 1.7%–25.1% increase in rainfall in the wet months. This will translate to frequent and more intense flooding and dry spell events, resulting in recurring crop damage and worsening poverty.

The increasing frequency of inter-annual anomalies from both typhoons and droughts in recent years foretell the impacts of global warming. In the early half of 2010, prolonged dry spells in the Philippines caused rice and corn crops to wilt. In November of the same year, Typhoon Juan hit Isabela Province, inflicting Philippines Peso (PHP) 542 million in damages on Ilagan City, with 45% of the total damages attributed to the agricultural sector. The following year, Typhoon Quiel slammed Isabela, causing PHP 115 million in damages, and displacing more than 500,000 people (NDRRMC 2011). In the Abuan floodplain, some 2342 residents from seven *barangays* (villages) were evacuated and were housed in temporary shelters due to the flooding of those communities (DRRMO, City Government of Ilagan (NDRRMC 2011).

Ilagan City lies at the confluence of the Cagayan and Ilagan rivers. It has a population of 33,000 households in 91 *barangays*, with majority of the residents being corn and rice farmers. The economic base of the city is agriculture, with 28,000 hectares (ha) of cornland and 7000 ha of rice land. Other crops planted are sugar cane, tobacco, and cassava. On the eastern *barangays* of the city lies the Abuan watershed, a 63,754 ha ecosystem that supports some 2900 farming households. The upper catchment has an area of 44,000 ha and is located in the Northern Sierra Madre Natural Park, which supports the last remaining old-growth dipterocarp forests in the country. The lower sub-catchments consist of farmlands, residual forests, and brushland, with an area of 19,000 ha (or 31%) of the watershed area. The watershed is named after the Abuan River, which, together with the smaller Bintacan River, drains into the Ilagan River before merging with the Cagayan River (Fig. 3.1). Flooding in the Abuan floodplain is triggered when flows are impeded when the Ilagan River and Cagayan River break their banks and inundate the delta. The figure shows the map of the Abuan watershed. The lower Abuan catchments (in yellow) consist of tenured lands with an area of 4057 ha.

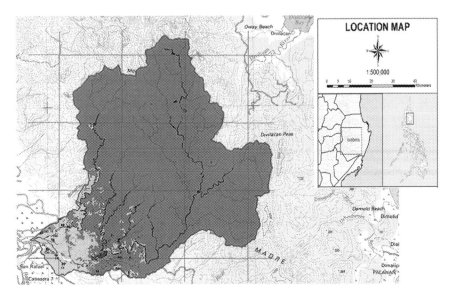

Fig. 3.1 Map of study area

3.3   Implementation Strategies and Activities

Recognizing the importance of LGUs under a devolved setup, the AIWMP worked at three levels to produce policy outcomes for CCA and DRRM in the City of Ilagan:

(a) The AIWMP implemented innovative activities at the barangay level through the Barangay DRRM councils in coordination with City DRRMO. The AIWMP trained barangay and city officials on the use of tools and state-of-the-art technologies, i.e. GIS maps, crop models, weather and flood sensors as decision support system for farmers and government.

(b) The city government officials, convinced of the effectiveness of the demonstration activities in the pilot barangays, became interested in the widespread use of these new tools and practices to other city areas.

(c) The AIWMP assisted the LGU in DRRM and CCA policy and made preparations as required by the law. Light Detection and Raging (LIDAR) maps and Community-Based Monitoring Systems (CBMS) were made available to local residents, which allowed the AIWMP to demonstrate their findings to DRRM and CCA policy formulation, and to subsequently integrate them into the LCCAP of the city.

3.3.1   Flood Risk Mitigation Activities

3.3.1.1   Installation of Automated Weather Stations

The project installed rain gauges and water-level sensors at bridges along the Abuan River and Bintacan River as part of establishing an Early Flood Warning Systems. A Memorandum of Agreement (MOA) with the Department of Science and Technology (DOST) enabled the project team to include these stations in Project NOAH (Nationwide Operational Assessment of Hazards), the Philippine's flagship program on disaster prevention and mitigation, and to be allowed access to data from its eastern stations in Isabela, where most storms originate. The contribution of the Ilagan River to flooding in the Abuan floodplain is considerable.

3.3.1.2   Topographic Mapping and Flood Simulation

The project completed topographic and bathymetric surveys using unmanned aerial vehicles (UAVs). A digital elevation model and hydrologic data were produced and inputted into a hydrologic model to simulate flooding in the Abuan floodplain. Flood inundation maps for 5-, 25-, and 100-year storm return periods were produced. Figure 3.2 shows a 100-year return period. Flood depths of 2.5 m (red areas) were found to overlap with human settlements threatening households. Flood models were generated using HEC-HMS, HEC-RAS and ARC-INFO.

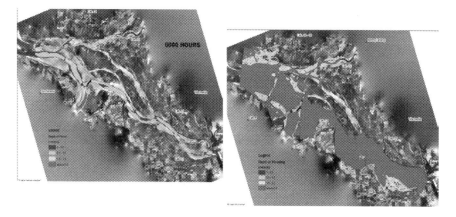

Fig. 3.2 Development of flood maps: before-and-after flood risk map of Abuan flood plain

3.3.1.3   Vulnerability Mapping and BDRRM Planning

Use of digital spatial mapping was introduced to analyze climate hazard vulnerabilities in community study areas. The flood inundation maps that had been produced were presented in Abuan barangays for community validation. This was followed by *barangay* training and planning workshops. Through the barangay focus group discussions, the vulnerable groups (i.e., flood-prone *sitios*,[1] elderly, persons with disabilities, women, and children) were identified by the community. Accordingly, *barangay* plans were prepared, which included the establishment of early warning systems (EWS), evacuation sites and camp management plans. The community also identified schools and covered courts as future evacuation centers. Should these places prove to be unfeasible, the *barangay* LGU identified elevated areas that could be used as temporary staging areas.

3.3.1.4   EWS

The City DRRMO established coordination protocols for *barangay* officials in issuing typhoon warnings to residents. As part of the established EWS, *barangay* officials are now using SMS, two-way radios, light-emitting diodes boards, and megaphones to issue alerts. Following the practice in the province, forced evacuation is also being implemented in flood-prone *barangays* whenever a typhoon enters Isabela. In a typhoon alert, the City DRRMO monitors the rainfall and water level readings via internet (www.fmon.asti.dost.gov.ph). The limitation of this service, however, is it depends on power and an internet connection, which may not function during a typhoon.

3.3.2   Drought Risk Mitigation Activities

3.3.2.1   Crop Modelling for Corn Production and Climate Change Impact Assessment

Crop models enable scientists to understand and to predict crop responses to global warming, as indicated by increasing CO_2, temperatures, and water scarcity. Accordingly, the project installed an agro-meteorological station and soil moisture sensors in the Abuan watershed to record rainfall, temperature, and other parameters. Two crop models (i.e., Decision Support System for Agri-Technological Transfer [DSSATT] and AquaCrop) were calibrated and validated within reasonable accuracy during two cropping seasons (Tongson et al. 2017; Balderama et al. 2017). The crop model simulated the impacts of climate change on biomass and yield. The model can also simulate present impacts, which can provide valuable

---

[1] A *sitio* is a territorial enclave that forms part of a *barangay*.

insights for farmers and help them to make informed decisions regarding the day-to-day activities on their farms.

3.3.2.2 Text Messages of Daily Weather Forecasts

The DSSAT crop model was automated to generate weather and crop advisories to farmers via SMS. Daily weather forecasts were made available using a numerical weather forecasting model called Weather Information-Integration for System Enhancement (WISE), which was developed by IBM Philippines and the Institute of Environmental Science and Meteorology (IESM) of the University of Philippines under Project NOAH. These daily forecasts can be viewed on the internet (www.weather-manila.com) or at the Project NOAH website.

The proof of concept for the weather and crop advisories was developed in 2016–2017 (Trogo et al. 2015a, b; Ebardaloza et al. 2015). The system was then tested to a pilot group of 30 farmers using GSM cellular phones. Each of the pilot farmers was given a prepaid SIM card, and was tasked to enroll their farms, send SMS queries, and receive replies from the system. The system can respond to queries about daily rainfall, amount of fertilizers required, best planting dates, yields, etc. The system then processes each query using these parameters, and responds to the farmer. The SMS queries were tallied and categorized during the pilot implementation. Almost all of the queries tallied were about rainfall forecasts, indicating the importance of this parameter to farmers. The list expanded to more than 1000 farmers as the LGU enrolled more farmers to receive text messages from the system.

3.3.3 Policy-Level Activities

3.3.3.1 Vulnerability Assessments and Mapping

The DILG introduced the CBMS as a tool that LGUs can use to target households, for bottom-up planning, and budgeting (DILG-MC 2014). Accordingly, the CBMS of Ilagan City consists of a database of 33,000 geo-tagged households and includes indicators relating to income, education, housing, access to basic services, water, sanitation, flood risks, etc. that are generated at the household and individual levels.

The city government then obtained copies of high-resolution (0.5 m) LIDAR maps and flood inundation maps from the Disaster Risk and Exposure Assessment for Mitigation (DREAM) Project of the DOST. The AIWMP overlaid flood inundation maps (Fig. 3.3) with CBMS household maps to generate maps of vulnerable households (i.e., low income, lacking access to safe water, lacking access to toilets, people living in makeshift shelters, and people with informal tenure), and maps of corn and rice growing areas. Figure 3.4 shows a vulnerability map of central Ilagan City overlain with households classified as poor and not poor.

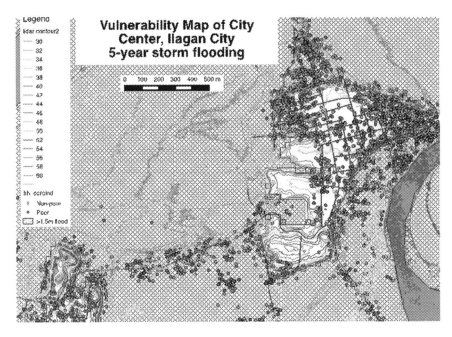

Fig. 3.3 Flood vulnerability map of central Ilagan City

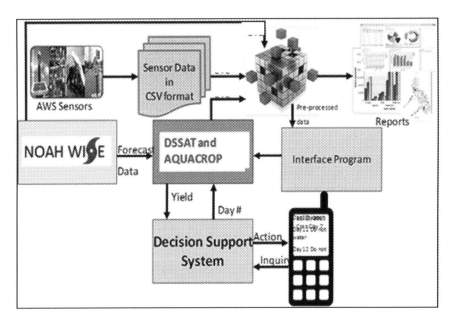

Fig. 3.4 Architecture design of decision support system

3.3.3.2   Downscaling Global Circulation Models and Impact Modelling

Assessing the impacts of climate change primarily requires understanding past trends in local climate (baseline) and projecting future trends based on best-fitting climate models. Secondly, climate projections are assessed against possible impacts on agriculture, water supplies, and health. This requires expertise in performing impact assessments for particular sectors of interest. In this case story, impact assessments were produced for corn and water resources.

The CSIRO-k3 (Commonwealth Scientific and Industrial Research Organization) GCM (Global Circulation Model) (Gordon et al. 2002) was selected to generate climate projections for 2050s and 2090s for Ilagan City. Daily weather files were generated using the MarkSim weather generator, and the data were inputted into crop models (Jones and Thornton 2000). The projected climates for 2050s and 2090s were used to simulate corn production (Tongson et al. 2017) and the changes to water resources in the Abuan watershed (Rojas 2017). Findings showed that corn yields would decrease up to 60%, and water loss due to evaporation would increase up to 33%, reducing the amount of water for economic uses.

3.3.3.3   Formulation of LCCAP Chapter 2

Chapter 2 of the LCCAP consists of climate projections, risk exposure of vulnerable groups to climate change and the corresponding policies to address them, adaptive capacity maps, and CCA programs and plans. The LCCAP was adopted by the LGU through a resolution passed by the City Council.

3.3.4   Development of Farmers Decision Support System

The Farmer Decision Support System (FDSS) is an ICT-enabled internet and SMS platform capable of advising individual farmers on how crops will respond given the climate forecast for the next cropping season. It is also a diagnostic tool that helps farmers to identify yield gaps and their sources. The model inputs are data from the climate seasonal forecast of the NOAH-WISE, five-day weather forecast from the Weather Data Solutions of IBM (IBM-WEDA), and actual daily weather from the weather stations of the Advanced Science and Technology Institute of the DOST (DOST-ASTI) (Trogo et al. 2015a, b). Based on inputs on weather, soil, and farmer practices, the platform simulates crop responses through the DSSAT crop model for corn (Balderama et al. 2017). The outputs of the model runs are crop calendar, biomass, and yields, which are sent by SMS or downloaded over the internet.

3.4 Lessons Learned

1. The local agricultural landscape in Abuan watershed consists of a mosaic of smallholder farms characterized by high landscape variability and heterogeneity. This study provided a clear understanding of climate change, the El Niño-Southern Oscillation, and increasing intensity of natural disasters that adds to the complexity of issues that are faced by small farmers. The other realities highlighted by this study are the ineffective agriculture extension services, which are constrained by lack of capacity and resources. Hence, these smallholder farmers are facing an uphill battle against poverty, degraded ecosystems, and worsening climate change.

2. From the project strategies implemented, the bottom-up demonstration activities and policy level activities resulted in the following policy outcomes:

   (a) *Development of Sangguniang Panlungsod Resolution (Provincial Council Resolution) No. 038.* Resolution approving the climate change action plan of Ilagan City for CY 2017–2022. This provides an effective strategy for mainstreaming research and development outputs into policies that can be implemented by the local governments of the province.

   (b) *City Ordinance No. 117–2016.* On a municipal level, an ordinance implementing flood evacuation drills in all flood-prone *barangays* in Ilagan City on a regular basis to increase the *barangays'* capacity and to ensure that the communities are prepared in cases of disasters and/or calamities. This activity can now be easily rolled-out to other localities, given a legal mandate and actual demonstration showcase.

3. Looking at the current legal framework used in the Philippines for CCA and DRRM, which has involved extensive stakeholder consultations, a suitable method has been found through which the legal framework can provide entry points for projects like AIWMP to embed its activities within the framework of LGU plans and programs. This study provided a good example of how to align and plan the interface of a project framework to involve local governments and facilitate the effective adaption among stakeholders.

4. It was also concluded that while technologies and innovations introduced by the project show great potential and offer opportunities for modernizing agriculture, farmers and city extension workers' knowledge of crop models, remote sensing, and their potential use in decision making are still rudimentary and still needs more field validations. This could be made part of future training programs and other future endeavors to improve and sustain initial success of the project

5. It was also determined that access to ICT is also expensive in rural areas, and thus constrains use of state-of-the-art technology. There is therefore a need for a national broadband policy and ICT investments in underserved areas. The pilot introduction of information and communication technologies-enabled (ICT-

enabled) crop and weather advisories have demonstrated great promise for modernizing agriculture; however, such a feat would require an enabling policy environment in the ICT sector in order to stimulate growth and benefit the agriculture sector.

6. Another important lesson is mainstreaming of multidisciplinary approaches to CCA and DRRM drawing from natural, social, and ICT fields that were facilitated at the local level. Parallel interventions at the national level (involving the Department of Information and Communications Technology, Department of Agriculture, Department of Science and Technology, state universities and colleges, and the private sector) will require an enabling policy environment.

7. Finally, the AIWMP demonstrated use of innovative ICT-enabled technologies that can help to modernize the agricultural sector. Precision agriculture and individualized advice to farmers using remote sensing, SMS, and internet show great promise. However, there is a need for more capacity building among LGU agriculture workers in ICT and crop modeling. An enabling policy environment in the ICT sector will spur development in applications that benefit the agricultural sector and small farmers.

References

Balderama O, Alejo L, Tongson E, Pantola RR (2017) Development and application of corn model for climate change impact assessment and decision support system: enabling Philippine farmers to adapt to climate variability. In: Leahl Filho W (ed) Climate change research at universities. Hamburg University, Springer, pp 373–387

DILG-MC (Department of Interior and Local Government Memorandum Circular) 2014-135. Guidelines on the Formulation of Local Climate Change Action Plan (LCCAP). DILG. (21 October 2014)

Ebardaloza JBR, Trogo R, Sabido DJ, Tongson E, Bagtasa G, Balderama OF (2015) Enabling Philippine farmers to adapt to climate variability using seasonal climate and weather forecast with a crop simulation model in an SMS-based Farmer Decision Support System. Accessed 7 Oct 2017. http://adsabs.harvard.edu/abs/2015AGUFMGC53G1298E

Gordon HB, Rotstayn LD, McGregor JL, Dix MR, Kowalczyk EA, O'Farrell SP, Waterman LJ, Hirst AC, Wilson SG, Collier MA, Watterson IG, Elliot TI (2002). The CSIRO Mk3 Climate System Model. CSIRO Atmospheric Research Technical Paper No. 60. Aspendale. Commonwealth Scientific and Industrial Research Organisation, Victoria

Jones PG, Thornton PK (2000) MarkSim software to generate daily weather data for Latin America and Africa. Agron J 92(3):445–453

NDRRMC (National Disaster Risk Reduction and Management Council) (2011) NDRRMC SitRep No. 12: Effects of Typhoon 'QUIEL' (Nalgae). Accessed 4 April 2016. http://www.ndrrmc.gov.ph/attachments/article/1763/SitRep_No_12_effects_of_Typhoon_QUIEL_as_of_10OCT2011_0600H.pdf

Philippine Atmospheric Geophysical and Astronomical Services Administration (2011) Climate projection for Isabel Province, Quezon City

Rojas D (2017) Hydrology of Abuan watershed. Internal Report to Worldwide Fund for Nature. Quezon City Philippines

Tongson EE, Alejo L, Balderama OF (2017) Simulating impacts of El Niño and climate change on corn yield in Isabela, Philippines. Clim Disaster Dev J 2(1):29–39

Trogo R, Ebardaloza JB, Sabido DJ, Bagtasa G, Tongson E, Balderama OF (2015a) SMS-Based Smarter Agriculture Decision Support System for Yellow Corn Farmers in Isabela. In: Proceedings of the 2015 IEEE Canada International Humanitarian Technology Conference (IHTC2015), 31 May 2015 to 4 June 2015, Ottawa, Canada. Ottawa, Canada: Institute of Electrical and Electronics Engineers

Trogo R, Ebardaloza JB, Sabido J, Tongson E, Bagtasav G, Balderama OF (2015b) Enabling Philippine farmers to adapt to climate variability using seasonal climate and weather forecast with a crop simulation model in an SMS-based Farmer Decision Support System. Paper presented at the 2015 AGU Fall Meeting, San Francisco, CA, 14–18 December 2015

Chapter 4
Economic Evaluation and Climate Change Adaptation Measures for Rice Production in Vietnam Using a Supply and Demand Model: Special Emphasis on the Mekong River Delta Region in Vietnam

Yuki Ishikawa-Ishiwata and Jun Furuya

4.1 Background

Of the countries most affected by the impact of climate change between 1999 and 2018, Vietnam ranked sixth when factors such as monetary losses, mortality, and frequency of extraordinary climatic events are considered (Eckstein et al. 2020). The agricultural sector is one of the most susceptible to climate change because crop production is directly influenced by climate conditions. Rice is the staple food in Vietnam and, as of 2018, 65.3% of crop production in the country was dedicated to rice cultivation (General Statistical Office of Vietnam, GSO 2020). Flooding and saline intrusion due to sea level rise (SLR), as well as droughts associated with El Niño events, have caused serious problems in the country in the past, and these events have the potential to adversely affect rice production in the future (Lasco et al. 2011; Nhung et al. 2019). Surplus rice can be exported and sold on the global rice market, and Vietnam was the second largest rice exporter and fifth largest producer in 2020 (United States Department of Agriculture, Production, Supply and Distribution, USDA PS&D 2020). Consequently, the effects of climate change on rice production in Vietnam can affect the global rice market. Measures such as shifting transplanting dates, introducing irrigation systems, improving irrigation efficiency, using short-duration and salinity- or thermo-tolerant cultivars have been adopted as adaptation measures to tackle climate change (Masumoto and Tada 2008; Yu et al. 2010, 2013; Shrestha et al. 2014; Trinh et al. 2014). Adaptation is

Y. Ishikawa-Ishiwata (✉)
Global and Local Environment Co-creation Institute, Ibaraki University, Ibaraki, Japan
e-mail: yuki.ishikawa.ga@vc.ibaraki.ac.jp

J. Furuya
Social Sciences Division, Japan International Research Center for Agricultural Sciences, Ibaraki, Japan

© The Author(s) 2022
T. Ito et al. (eds.), *Interlocal Adaptations to Climate Change in East and Southeast Asia*, SpringerBriefs in Climate Studies,
https://doi.org/10.1007/978-3-030-81207-2_4

complicated by the fact that the effects of climate differ by region; for example, saline water intrusion, drought, flooding, and coastal erosion are the main issues affecting the Mekong River Delta region, while drought is a serious problem in the northern mountainous areas (Trinh et al. 2014). Thus, the development of adaptation measures needs to consider these differences and target specific regions. Investigating the effect of climate change on rice production and the rice market after the introduction of appropriate adaptation measures in each region can be evaluated economically using a supply and demand model (Furuya and Meyer 2008).

The supply and demand model considers yield, planted area, exports, imports, stock changes, and food demand functions of rice. From these functions, supply and demand equilibrium prices can be estimated and the outlook of future rice production and equilibrium prices can be assessed. Inserting climate variables, such as temperature or evapotranspiration into yield and planted area functions, facilitates projections of the effect of climate change on rice production and the market. With these projections as a baseline, rice production and equilibrium prices can then be estimated, and adaptation measures can be implemented based on different scenarios. These methods facilitate analyses based on quantitative values and allow managers to consider adaptation measures that are appropriate for each region. In the next section, we will introduce the concept of the supply and demand model for rice in Vietnam. In Sect. 3, we will demonstrate the economic evaluation methods used to develop adaptation measures in response to climate change using this model. In Sect. 4, we will examine extant adaptation measures and problems for farmers in the Mekong River Delta, which is the main agricultural production region in Vietnam. Finally, we will consider future research prospects.

4.2 Supply and Demand Model with Climate Variables

Market equilibrium is the basic concept of the supply and demand model. The relationship between price (P) and quantity (Q) is shown in Fig. 4.1. The intersection between the supply and demand curves is the equilibrium price. If production is reduced due to climate change, then the price will increase, all things being equal. On the other hand, if the personal income of consumers increases and consumption increases, then prices will increase. The equilibrium price is thus determined based on a balance between the production of farmers (supply) and consumption by consumers (demand). The supply curve can be shifted through technological advancement or large-scale agricultural development (Fig. 4.1), and these would cause the price to decline. In addition to consumer behavior, price reduction can also be affected by trends in exports and/or imports to/from other countries. Thus, the influence of these factors agricultural markets can be assessed using the supply and demand model. In this model, the planted area function with farm prices as explanatory variables can be regarded as a supply function, and the food demand function with farm prices can be regarded as a demand function. The equilibrium prices can be estimated by convergence using the Gauss-Seidel iterative method. By

Fig. 4.1 Concept of supply and demand curve and equilibrium price

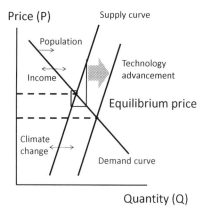

simulating the equilibrium price with historical value, medium- and long-term projections of crop production, export, import, and food demand can be estimated (Furuya and Meyer 2008; Furuya 2020).

The flowchart in Fig. 4.2 shows the relationships between supply and demand and climate variables, such as evapotranspiration. Furuya et al. (2010) described the functions and explanatory variables used in the model. The supply and demand rice model can be applied to a country by dividing rice production, which is estimated based on yield and planted area, into distinct regions within that country. When the model is applied to the world, then the supply and demand model needs to consider the main producing and consuming countries and how they are linked to individual countries (see Ishikawa-Ishiwata and Furuya (2021a) for a description of the global supply and demand model).

In the supply and demand rice model, income (GDP) and population can be treated as exogenous variables and rice consumption per capita (kg/person) can be estimated. The amount of rice production (metric tons, MT) can be determined from yield (MT/ha) and planted area (ha). At least 20 years of historical data are needed for the analysis because the model is a historical statistical model. Historical yield, planted area, stock changes, exports and imports, food demand, domestic farm prices, GDP, GDP deflator, population, and the exchange rate can all be obtained from the USDA and Food and Agricultural Organization Corporate Statistical Databases (FAO-STAT). The consumer price index (CPI) can be obtained from the World Development Indicator (WDI) and the global price of rice can be obtained from the World Bank (WB). Provincial yield and planted area can be obtained by GSO.

By measuring the yield and planted area function with climate variables, such as evapotranspiration, the effect of climate change on rice production can be estimated. The solution derived from the model is the baseline. Figure 4.3 indicates the baseline results estimated by the supply and demand model. Evaluations considering climate change adaptation measures can be conducted by incorporating the scenario data into the baseline simulation. Of the many measures that have been proposed to

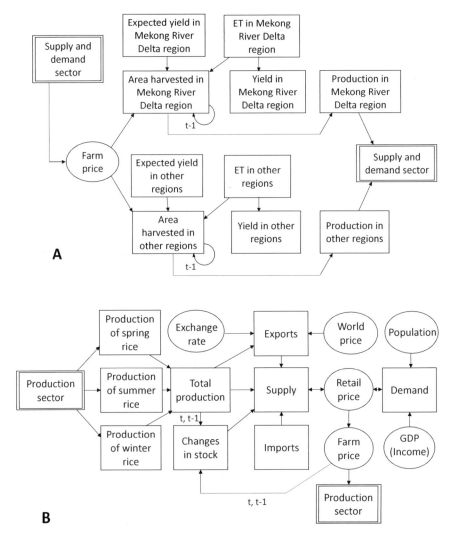

Fig. 4.2 Flowcharts of the rice production sector (**a**) and supply and demand sector (**b**) in the Vietnamese rice model. ET: evapotranspiration. Source: Furuya et al. (2010) with permission of the publisher

mitigate climate change, changing the cultivars is the most practical solution for farmers (Lasco et al. 2011). Here, we will explain the economic evaluation method used to assess the introduction of new cultivars as an adaptation measure.

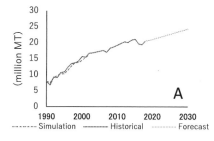
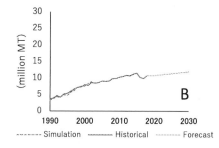

Fig. 4.3 Production of spring rice in Vietnam (**a**) and Mekong River Delta (**b**).
Source: updated Furuya et al. (2010)

4.3   Economic Evaluation of Climate Change Adaptation Measures Using a Supply and Demand Model for Rice

Vietnam rice production has been boosted through the adoption of high-yield culti-vars (Kien et al. 2020). Consequently, short duration and salinity- or thermo-tolerant cultivars are considered to play an important role in maintaining the stability of rice production within the context of climate change adaptation measures (Trinh et al. 2014). Detailed functions and an economic evaluation are described in Ishikawa-Ishiwata and Furuya (2020) and Ishikawa-Ishiwata and Furuya (2021a, b), who examined the introduction of new cultivars with a supply and demand model. Here, we describe the scenarios for the introduction of new cultivars. First, detailed his-torical data should be collected for the region of interest because the local environ-ment should be considered to determine whether the new cultivar has been accepted by local consumers (e.g., factors affecting changes in the taste or appearance). Based on information on the background of why the new cultivar is needed in the region, the year it was developed, time-series changes in the planted area of the cultivar, and changes in production costs (especially the cost of fertilizer and/or pesticide for the new cultivar), a new technology dissemination model (Griliches 1957) is incorporated into the scenario. The reason why farmers require the new cultivars (e.g., adaptation to high temperatures and saline water intrusion) should also be incorporated into scenario projections to provide robust quantitative fore-casts of crop production within the context of climate change. In the case of yield projection at higher temperatures, inserting the crop model into the yield function is effective. However, when considering the intrusion of saline water, actual damage should be investigated by field surveys in the scenario setting.

An economic evaluation for the introduction of the new cultivar should then be conducted based on a survey of the production costs. Comparing the planted areas of the new and previous cultivars, economic evaluations of the introduction of the new cultivar can be performed by comparing the production cost from the projec-tion of scenarios.

4.4 Actual Farmer Adaptation Measures in the Mekong River Delta

We have described the economic evaluation of adaptation measures for climate change by considering supply and demand. However, since rice farmers in the Mekong River Delta have been changing their farming systems, these shifts should be considered by the model. In the planted area function for rice in the supply and demand model, lagged planted area, farm price of rice and/or substitute goods need to be considered (Ishikawa-Ishiwata and Furuya 2020). In addition, climate variables, such as evapotranspiration, have been used as explanatory variables (Furuya et al. 2010). In the case of the Mekong River Delta, the farming system is complicated, and hence appropriate substitute goods should be considered. Here, we divided the delta into two major regions, the coastal to central regions of the Mekong River Delta, and the backswamp in the Mekong River Delta areas based on the farming systems used in these regions.

In coastal areas, the main type of farming is shrimp aquaculture in brackish water, and rice is not cultivated. In central regions of the Mekong River Delta, the main type of farming is rice farming-shrimp aquaculture using brackish water, and farmers utilize the saline water that has intruded into this region during the dry period effectively. During the rainy season, this saline water is flushed from the aquifer and sediments by rain water, which enables farmers to cultivate rice in the rainy season. In the backswamp regions of the Mekong River Delta, double- and triple-cropping of rice have been conducted (Japan International Cooperation Agency, JICA 2013). Additionally, climate change issues differ between regions; saline water intrusion during the dry season occurs in the coastal to the middle parts of the Mekong River Delta and flooding occurs during the rainy season in the backswamp areas of the Mekong River (JICA 2013). Consequently, the main countermeasures enacted by the government differ between areas (Table 4.1). Constructing dikes and introducing short-duration cultivars can make it possible for rice farmers to perform double- and triple-cropping of rice in the flood-prone backswamp areas,

Table 4.1 Problems facing rice farmers in the Mekong River Delta and farmers' adaptation measures. Source: Cuc et al. (2008); JICA (2013); Paik et al. (2020)

Main problems	Government adaptation measures	Farmers' adaptation measures	Region
Saline water intrusion	Construction of sluice gates	- Shifting crop farming to aquaculture-based farming. - Introducing salinity tolerant crop cultivars	Coastal to central areas of the Mekong River Delta
Flood	Construction of polder-dikes	- Adopting short-duration crop cultivars to enable early harvest. - Optimizing the crop calendar according to the inundation period	Backswamp areas in the Mekong River Delta

such as in An Giang Province (Fig. 4.4), one of the top rice-producing provinces in the Mekong River Delta; these measures have boosted the rice production in Vietnam (Kien et al. 2020). According to local people, the construction of dikes and the subsequent alleviation of floods allowed farmers to increase the input of fertilizers.

However, while this increase in fertilizer input has enabled local farmers to perform triple-cropping, the soil quality has deteriorated and the amount of fertilizer required to sustain these levels of productivity has increased annually (Nguyen et al. 2018; Tran et al. 2018). In addition to the increased input of fertilizers, pesticide application increased three- to six-fold during 2000–2015 and this has had the effect of increasing production costs (Kien et al. 2020). The low price of Vietnamese rice is a key factor explaining its dominance on the global rice market, and increases in production costs combined with increases in fertilizer input will make it difficult for the country to maintain its global dominance. Such considerations are not only important for export from Vietnam, but also for the importing countries. In addition, increased rice production costs would adversely affect rice farmers, who may turn to aquaculture if their income decreases. Thus, the area under rice cultivation may decrease and this could lead to a reduction in the production of rice in Vietnam that is in excess of the possible reduction in yield due to climate change.

In general, official publicly available statistical data, such as FAO-STAT, USDA, and GSO data, are used in the supply and demand model. Due to the different types of farming systems employed in the Mekong River Delta, application of the model should be carefully considered. For example, according to local people in Soc Trang Province, which extends from the coast to the central region of the Mekong River Delta, mono-cropping of rice has shifted to shrimp aquaculture in coastal areas and double-cropping of rice has shifted to triple-cropping in the inland areas. Although the area under rice cultivation has increased and decreased as styles of farming have changed, the planted area has been stable from 1995 to 2019 (Fig. 4.4). Based on a

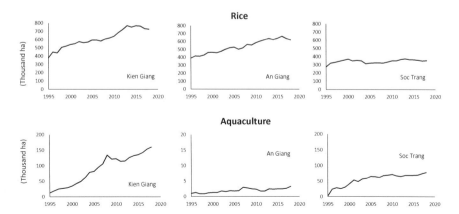

Fig. 4.4 Annual rice (upper row) and aquaculture (lower row) production in Kien Giang, An Giang, and Soc Trang provinces. Source: GSO. *Note: the vertical axis for aquaculture in An Giang Province differs from that for the other provinces*

report published by the Sub-National Institute of Agricultural Planning and Projection (Sub-NIAPP), extensive areas of rice paddies will be converted to brackish water aquaculture due to saline intrusion by 2050 (JICA 2013). This implies that the area under rice cultivation will be reduced significantly in the future. Since the selection of farming systems by farmers varies and affected by profit, competition or changing farming systems based on profitability and risks should be considered by managers when allocating areas to rice cultivation using the supply and demand model in the future.

4.5   Lessons Learned

The climate change adaptation measures required for continued rice production in Vietnam have been investigated in terms of technology, such as by adopting short-duration or salinity-tolerant cultivars. However, farmers not only adopted new technology but also changed the farming systems. For example, rice farmers have shifted to aquaculture-based farming in the coastal and central regions of the Mekong River Delta. Farming systems are expected to change further due to saline intrusion caused by climate change in the future. Considering the actual farmers' situation and a variety of scenarios, practicable evaluations can be conducted in each region. The effect of the area under cultivation on the profitability of different farming systems can be complemented with more practical simulations in the future.

References

Cuc NH, Cruz PCS, Borromeo TH, Hernandez JE, Tin HQ (2008) Rice seed supply systems and production in Mekong Delta, Vietnam. Asia Life Sci 17:1–20
Eckstein D, Künzel V, Schäfer L, Winges M (2020) Global climate risk index 2020. Who suffers most from extreme weather events? Weather-related loss events in 2018 and 1999 to 2018. Germanwatch
FAO Statistical Databases (FAO-STAT). http://www.fao.org/faostat/en/#home
Furuya J (2020) Development of an economic model for evaluation of climate change in the long-run for international agriculture: EMELIA. JIRCAS Working Report 89
Furuya J, Meyer SD (2008) Impacts of water cycle changes on the rice market in Cambodia: stochastic supply and demand model analysis. Paddy Water Environ 6:139–151
Furuya J, Meyer SD, Kagyama M, Jin S (2010) Development of supply and demand models of rice in Lower Mekong River Basin Countries: REMEW-Mekong. JIRCAS Working Report 68
General Statistics Office of Vietnam (GSO) (2020). https://www.gso.gov.vn/Default_en.aspx?tabid=491
Griliches A (1957) Hybrid corn: an exploration in the economics of technological change. Econometrica 4:501–522
Ishikawa-Ishiwata Y, Furuya J (2020) Evaluating the contribution of soybean rust-resistant cultivars to soybean production and the soybean market in Brazil: a supply and demand model analysis. Sustainability 12:1422
Ishikawa-Ishiwata Y, Furuya J (2021a) Soybean rust and resistant cultivar effects on global soybean supply and demand. Jpn Agric Res Q 55(1):59–67

Ishikawa-Ishiwata Y, Furuya J (2021b) Fungicide cost reduction with soybean rust-resistant culti-
   vars in Paraguay: a supply and demand model approach. Sustainability 13:887
Japan International Cooperation Agency (JICA) (2013) The project for climate change adaptation
   for sustainable agriculture and rural development in the coastal Mekong Delta in Vietnam.
   Final report: priority project
Kien NV, Han NH, Cramb R (2020) Trends in rice-based farming systems in the Mekong Delta. In:
   Cramb R (ed) White gold: the commercialization of rice farming in the lower Mekong Basin,
   pp 347–374
Lasco RD, Habito CMD, Delfino RJP, Pulhin FB, Concepcion RN (2011) Climate change adaptation for smallholder farmers in Southeast Asia. World Agroforestry Centre, Los Baños
Masumoto T, Tada M (eds) (2008) The assessment of changes in water cycle on food production
   and alternative scenarios. –Implications for policy making. ISBN 978-4-9902838-7-2 C3061.
   (in Japanese)
Nguyen VK, Dumaresq D, Pittock J (2018) Impacts of rice intensification on rural households in
   the Mekong Delta: emerging relationships between agricultural production, wild food supply
   and food consumption. Food Sec 10:1615–1629
Nhung TT, Vo PL, Nghi VV, Bang HQ (2019) Salt intrusion adaptation measures for sustainable
   agricultural development under climate change effects: a case of Ca Mau peninsula, Vietnam.
   Clim Risk Manag 23:88–100
Paik SY, Le DTP, Nhu LT, Mills BF (2020) Salt-tolerant rice variety adoption in the Mekong river
   Delta: farmer adaptation to sea-level rise. Plos-One 15:e0229464
Shrestha S, Deb P, Bui TTT (2014) Adaptation strategies for rice cultivation under climate change
   in Central Vietnam. Mitig Adapt Strateg Glob Chang
Tran DD, Halsema G, Hellegers PJGJ, Ludwig F, Wyatt A (2018) Questioning triple rice intensifi-
   cation on the Vietnamese Mekong delta floodplains: an environmental and economic analysis
   of current land-use trends and alternatives. J Environ Manag 217:429–441
Trinh MV, Bo NV, Minh HG, Dzung NX (2014) Climate change and impacts on rice production
   in Vietnam: pilot testing of potential adaptation and mitigation measures. Deliverable 1.2 a
   benchmark report characterizing the three project areas and rice farming systems in the three
   provinces. Bioforsk:1–39
United States Department of Agriculture. Production, Supply, and Distribution. (USDA PS&D)
   (2020). https://apps.fas.usda.gov/psdonline/app/index.html#/app/advQuery
World Bank (WB). http://www.worldbank.org/
World   Develop   Indicators   (WDI).   https://databank.worldbank.org/source/
   world-development-indicators
Yu B, Zhu T, Breisinger C, Hai NM (2010) Impacts of climate change on agriculture and policy
   options for adaptation. The case of Vietnam. IFPRI Discussion Paper, 01015
Yu B, Zhu T, Breisinger C, Hai NM (2013) How are farmers adapting to climate change in Vietnam?
   Endogeneity and sample selection in a rice yield model. IFPRI Discussion Paper, 01248

Chapter 5
Small Coastal Island Ecosystems and Conservation Perspectives Within Adaptation Efforts

Dietriech Geoffrey Bengen

5.1 Introduction

Indonesia, which consists of 16,671 named islands and 104,000 km of coastline, is the biggest archipelagic state and a host to precious and essential coastal and small island goods and services.

While perceiving the extensive mangrove forests and vast coral reefs along the coasts of more than 10,000 small scattered islands from Aceh to Papua of the Indonesian archipelago, how many people stop appreciating the role of those ecosystems in human lives? These two primary coastal-small island ecosystems highlight strategic values for human lives. Mangroves and coral reefs perform four essential functions: (1) provide useful natural resources, (2) support essential services for human livelihood, (3) maintain leisure services, and (4) protect coastal areas from natural disasters. Small coastal island ecosystems contain various natural resources, which are essential for human lives and daily livelihood. As a support system for critical human services, Small coastal island ecosystems provide a clean environment to support human activities, in addition to providing aesthetic tourism sites and protection from numerous natural disasters that threaten coastal and small island areas (Bengen 2020).

Thus, referring to the four integral ecosystem functions, coastal areas, and the thousands of small islands in Indonesia is challenging and promising. Small coastal islands should not merely be exploited for their natural resources; they should also be placed at the center of human welfare and different forms of utilization, e.g., ports and transportation hubs, marine-based industry, and community housing. A survey of research and literature showed that at least 85% of marine species are connected in some way to small coastal island ecosystems at some stage in their life

D. G. Bengen (✉)
Faculty of Fisheries and Marine Science, IPB University, Bogor, Indonesia
e-mail: dieter@indo.net.id

© The Author(s) 2022
T. Ito et al. (eds.), *Interlocal Adaptations to Climate Change in East and Southeast Asia*, SpringerBriefs in Climate Studies,
https://doi.org/10.1007/978-3-030-81207-2_5

cycle, and about 90% of fisheries yields originated from small coastal island ecosystems (FAO 2000).

Unfortunately, extensive damage has occurred to small coastal island ecosystems, not only threatening the natural capacity of the small coastal island in restoring their resources but reducing their capacity to mitigate natural disasters and exposing their vulnerabilities. Impacts include the loss of spawning habitats, degradation of nurseries and foraging areas for various species, and reductions in fish stocks. Further impacts include the loss of coastal ecosystems' physical functions, such as erosion prevention, wave reduction, seawater intrusion prevention, and waste absorption. Ecosystem damage originated from destructive practices associated with exploiting natural resources and nonenvironmentally friendly development in the small coastal island areas, e.g., blast fishing, extensive mangrove clearance, habitat alteration, and coral reef mining. Destruction of small coastal island ecosystems will lead to more adverse impacts when climate change impacts are considered. To restore valuable small coastal island ecosystem functions and services, particularly vital services for disaster mitigation, coastal-small island ecosystem protection and rehabilitation efforts should be undertaken in previously degraded areas.

5.2   Threats Related to Climate Change to Coastal-Small Island Ecosystems

As the interface between the terrestrial and marine ecosystems, small coastal islands are highly dynamic environments continually changing over relatively short periods. Typically, these dynamics are at equilibrium, but any significant change can negatively impact large and complicated ramifications. The increase in atmospheric greenhouse gas (GHG) emissions has changed the climate, warming air and sea surface temperatures, and increasing ocean acidification due to CO_2 diffusion into the ocean. Further heating may change rainfall patterns, causing sea level rise (SLR) and extreme weather events, like tropical cyclones (Anonymous 2008). Tangible consequences of climate change on small coastal island ecosystems include a rising sea surface, changes in seawater alkalinity, alterations in ocean circulation, localized upwelling due to vertical changes in seawater properties, sea-level rise, and increased frequency of extreme storm events, high tides, and heavy precipitation (Alongi 1998).

Regarding sea level rise, the last assessment by Working Group I of IPCC in February 2007 (IPCC 2007) showed that the sea level has risen by an average of 2.5 mm annually, and it was predicted to rise by 31 mm over the next decade. Indonesia, an island nation with over 10,000 small islands and 104,000 km of coastline, is very vulnerable to SLR. If SLR continues, Indonesia may lose as many as 2000 low-lying islands, including small coastal island ecosystems, by 2030 (Ministry of Environment of Republic of Indonesia 2007). Additionally, an SLR of between 8 and 30 cm by 2030 is likely to severely impact small coastal island ecosystems (PEACE, DFID, and World Bank 2007).

5.3 Importance of Established Conservation Areas for Adaptive Measures of Small Coastal Islands Towards Climate Change

There are responsive and planned adaptations to climate change. Responsive adaptations refer to changes in policy and behavior that people and organizations adopt in response to inevitable climate change and coastal and small island risks. Planned adaptation is intentional, proactive, and occurs at the societal level. It is more strategic than responsive adaptation and can address the full range of climate change hazards in coastal and small island resources in ways that meet social objectives.

Mainstreaming adaptation into the national development plan provides a strategy for planned adaptations in coastal and small island areas that highlighted actions that sought to "protect," "accommodate," and "retreat" from climate change impacts. While these approaches can be useful, coastal and marine adaptation has evolved to include a more integrated approach that incorporates ecosystem management and sustainable development.

The following types of adaptative actions could contribute to four outcome goals: (1) functioning and healthy coastal and marine ecosystems, (2) living shorelines, (3) reduced exposure and vulnerability of the built environment, and (4) strengthened capacity for adaptation.

In Indonesia, small coastal island areas include promising natural resources and environmental services, which are well suited to Indonesian sustainable marine development. The ever-increasing threat of climate change to small coastal islands, as previously described, intensifies ecosystem and natural resources degradation, including habitat degradation and biodiversity loss. Therefore, adaptive measures are required to mitigate climate change and protect ecologically and economically significant resources and maintain ecosystem integrity.

One adaptation to climate change impacts is establishing and developing coastal and small island conservation areas. Designated conservation areas should comprise intertidal to subtidal zones and the various associated flora and fauna with ecological, economic, social, and cultural values.

The main target of conservation in small coastal islands is the existing ecosystem and natural resources, and the aim is to maintain ecological processes, resource production, and environmental services for a sustainable human livelihood (Agardy 1997). Regarding the limited capacity of natural resources in small coastal island habitats, it is necessary to implement specific management schemes incorporating different types of users to ensure that the adaptive measures maintain these habitats' functional priorities. In this context, coastal-small island ecosystems bear intricate composition and dynamics between biophysical, social, and economic sub-system, which take place in the transitional landscape.

Small coastal island ecosystems are typically comprised of inter-related systems with particular landscape and time domains. Different traits of the biophysical, social, and economic systems in one landscape can affect one another. Thus, each spatial scheme may feature synergistic or conflicting interactions.

Ecosystem-based conservation management deals with managing different activities to utilize coastal-small island ecosystems. In the end, the increasing use of small coastal island resources should not exceed the natural functional capacity. Every natural ecosystem has four essential functions for humans: (1) providing useful natural resources, (2) support essential services for human livelihood, (3) maintain leisure services, and (4) protecting humans from hazardous natural disasters (Ortolano 1984). The first two functions are closely associated with the latter two functions. Thus, the natural ecosystem's second two functions keep working whenever the first two functions are preserved.

Within the framework of ecology, there are three prerequisites for accomplishing the optimum and sustainable management of small coastal island conservation areas: (1) spatial harmony, (2) assimilation capacity, and (3) sustainable use (Bengen 2002). Spatial harmony requires applying three different zones within one conservation area: preservation (no-take) zone, buffer zone, and the use zone (Fig. 5.1). Therefore, incorporating space within the conservation area should include a space protected from extraction activities (no-take zone). The ideal case for a preservation (no-take) zone would be for uses such as spawning grounds and coastal green belts, which usually serve as nurseries. In this preservation (no-take) zone, apart from scientific research and educational activities, other types of activities are prohibited. Different types of regulated exploitation activities, like fisheries and tourism, are permitted and limited exploitation activities in the use zone's buffer zone.

The preservation zone and the buffer zone in one conservation area are essential for maintaining different processes, which support sustainable resource production—for example, hydrologic and nutrient cycling, natural waste recycling, and biodiversity preservation. Furthermore, any exploitation activities (fisheries and mariculture) in the use zone should be managed to maintain biophysical harmony. Any exploitation activities in the utilization zone must consider several variables, such as (1) suitability of different uses to a particular existing ecosystem, (2) impacts associated with coastal activities and land-based activities (e.g., pollution, sedimentation, alternating hydrological regimes), as well as climate change-related impacts, and (3) compatibility between different utilization activities.

One of the small coastal island conservation strategies currently being developed as an adaptation measure in small coastal island ecosystems is a Marine Protected Area (Marine Sanctuary). The development of marine protected areas is done in order to maintain and improve the quality of small coastal island ecosystems and, at the same time, maintain and improve the quality of other marine living resources associated with the coastal-small island ecosystems. The objectives of marine protected areas are: (1) maintaining ecological functions by protecting habitats for living and spawning marine biota, and (2) maintaining the economic function of small

Fig. 5.1 Zonation of the small coastal island conservation area

coastal island ecosystems for local communities and their surroundings so that the sustainability of fishery production can be maintained and so that people's income will increase (Bengen et al. 2003; Faiza et al. 2010).

Figure 5.2 shows that the community's marine protected area (DPL) of Sebesi Island is divided into two zones: the core and buffer zones, in which the respective provisions apply. However, these activities are intended to protect marine resources, which the community will use to meet their daily needs (increased fish production). The prohibition on fishing gear in marine protected areas prevents damage to the small coastal island ecosystems and increases the fish resources associated with these small coastal island ecosystems. Utilization is only performed in a limited manner and uses simple non-destructive tools; further, it is carried out at a particular time, namely when the fishery has recovered. When the marine protected area has not been able to support the life of the coastal community of Sebesi Island, the social system of the Sebesi Island community seeks to find energy sources (alternative livelihoods) to meet their needs. It is done to prevent the removal of energy sources from the marine protected area. Therefore, some community members (system components) carry out marine cultivation activities or look for fish in places outside the marine protected areas to meet their daily needs (Bengen et al. 2003).

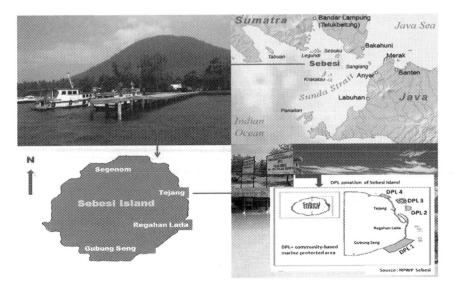

Fig. 5.2 Community-based marine protected area (DPL) of Sebesi Island, Lampung

As featured in a coastal-small island conservation area about various existing ecosystems, complexity indicates the urgency of integrated management in applied to corroborate adaptive measures to climate change impacts.

5.4 Lessons Learned

With the potential impact of climate change on the coastal-small islands, conservation areas' establishment and development are an adaptation strategy that protects small island ecosystems from damage due to climate change.

Small island coastal conservation measures require the assistance of both national and local governments, particularly in the following areas:

- technical assistance and funding mechanisms for local government in coastal resource management
- funding assistance criteria for local government
- incentives and disincentives
- commitments of both local and national government in sustainable coastal and marine resources management

Sustainable small coastal island conservation management requires thorough and participatory planning. It calls for coastal and marine resources management policies and strategies at national and local levels.

Prerequisites for small island coastal conservation management to be sustainable include initiatives that are:

- in concordance with formal and informal local policies
- inline with local communities' social and cultural conditions
- supported by human resources and institutional capacities
- have the involvement stakeholders
- a clear plan and program
- positive impacts on the environment, including local communities' socio-cultural and economic conditions.

For small island coastal conservation programs to have a widespread impact, careful consideration needs to be given to conservation replication design. The design requirement will include:

- a complete set of data and information (documentation process)
- evaluation of program relevance to conditions of particular areas
- evaluation of program impacts on the environment and communities
- involvement of stakeholders and communities in planning
- dissemination of information strategies.

Integrated small island coastal conservation management is one of the requirements to achieve optimal and sustainable management.

References

Agardy TS (1997) Marine protected areas and ocean conservation. Academic, San Diego

Alongi DM (1998) Coastal ecosystem processes. CRC Press LLC

Anonymous (2008) Adapting to coastal climate change. United States Agency for International Development

Bengen DG (2002) Coastal resources and ecosystems and its integrated and sustainable management. Marine Journalist Training Paper, organized by WWF-Wallacea Program, Bali, April 9–11, 2002

Bengen DG (2020) Integrated coastal zone management for coastal-small island ecosystem resilience in the face of climate crisis. Proceeding of the international workshop (Natural resource and risk management in the context of climate change: Southeast Asia Research-based Network on Climate Change Adaptation Science), Hanoi, January 9–10, 2020

Bengen DG, Tahir A, Wiryawan B (2003) Tinjauan Sustainabilitas, Akuntabilitas, Replikabilitas Pengembangan Daerah Perlindungan Laut di Pulau Sebesi, Lampung Selatan, Provinsi Lampung. Proyek Pesisir, Jakarta

Faiza R, Kusumastanto T, Bengen DG, Boer M, Yulianda F (2010) Sustainability of community-based marine protected area (case study: Blongko village, North Sulawesi, Sebesi Island, Lampung and Harapan Island, Jakarta). J Bijak dan Riset Sosek KP 5(1):19–30. (in Indonesian)

FAO (2000) The status of world fisheries and aquaculture. FAO Fisheries Department, Rome

IPCC (2007) Climate Change 2007: impacts, adaptation, and vulnerability. In: Parry ML, Canziani OF, Palutikof JP, van der Linden PJ, Hanson CE (eds) Contribution of Working Group II to the Fourth Assessment Report of the Intergovernmental Panel on Climate Change. Cambridge University Press, Cambridge

Ministry of Environment of Republic of Indonesia (2007) Climate variability and climate changes, and their adaptation. Ministry of Environment of the Republic of Indonesia, Jakarta

Ortolano L (1984) Environmental planning and decision making. Wiley, Toronto

PEACE, DFID, World Bank (2007) Indonesia and Climate Change status and policies. PEACE, DFID, and the World Bank

Part II
Disater Risk Reduction and Human Resource Development

Chapter 6
Geotechnical Approaches to Disaster Risk Reduction in Japan and Vietnam

Kazuya Yasuhara and Satoshi Murakami

6.1 Introduction

One of the factors responsible for the increased severity of disasters in recent years is overlapping events that are caused by a combination of climate change- and climate change-non-associated events. These compound disasters are common in Japan and Vietnam. This study attempts to explore ways in which to increase resilience against the risk of disasters, particularly in cases where residents encounter such compound disasters, in Japan and Vietnam.

6.2 Importance of Compound Disasters

Compound disasters require special attention because they magnify the extent of loss and damage. Figure 6.1a shows the characteristics of compound disasters. Typically, compound disasters have the following characteristics: (i) a second natural disaster occurs immediately before or after a major disaster, generating catastrophic consequences; (ii) damage is compounded through the combination of the natural disaster with a vulnerable natural background and/or human and social situations (see Fig. 6.1b); and (iii) the psychological aftermath amplifies the damage. This classification corresponds well with that proposed by Kokusho (2005).

K. Yasuhara (✉)
Global and Local Environment Co-creation Institute, Ibaraki University, Ibaraki, Japan
e-mail: kazuya.yasuhara.0927@vc.ibaraki.ac.jp

S. Murakami
Faculty of Engineering, Fukuoka University, Fukuoka, Japan

© The Author(s) 2022
T. Ito et al. (eds.), *Interlocal Adaptations to Climate Change in East and Southeast Asia*, SpringerBriefs in Climate Studies,
https://doi.org/10.1007/978-3-030-81207-2_6

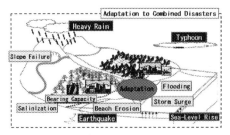 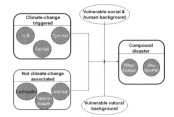

(a) Disasters are caused by multiple events

(b) Events and background associated with compound disasters

Fig. 6.1 Backgrounds of compound natural disasters (Yasuhara 2016) (**a**) Disasters are caused by multiple events (**b**) Events and background associated with compound disasters

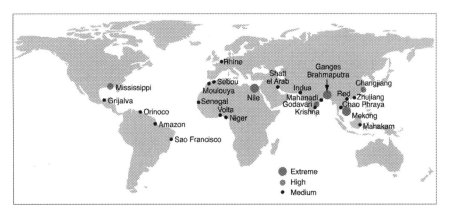

Fig. 6.2 Relative vulnerability of coastal deltas to sea-level rise (IPCC 2007)

6.2.1 Compound Disasters Related to Sea Level Rise (SLR)

This study describes the impacts of the compound effect of sea level rise (SLR) and land subsidence, which is a typical example of compound disaster related to climate change.

Figure 6.2 shows the location of major river deltas in the world, including the three largest deltas which are the Nile, Ganges, and Mekong river deltas (IPCC 2007). These river deltas have been designated in IPCC as extremely vulnerable coastal deltas.

Maruyama and Mimura (2010) conducted numerical analyses of the effects of SLR on inundation levels that can be expected at the end of the twenty-first century. Their results, which assume a scenario in which no adaptation measures have been put in place, showed that extensive areas in Asian regions will be inundated due to SLR.

The relative vulnerability shown in Fig. 6.3 does not consider the combined effect of SLR with land subsidence. To resolve this issue, the locations likely to

Fig. 6.3 Predicted areas (red) of inundation (after Maruyama and Mimura 2010; Mimura 2013)

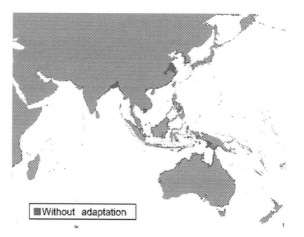

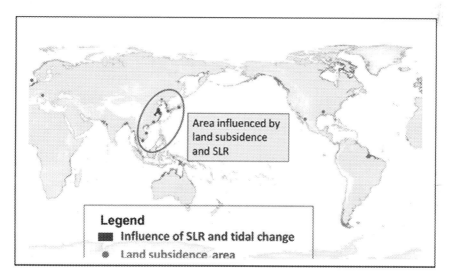

Fig. 6.4 Inundated areas obtained by combining SLR and tidal change data with land subsidence (from Maruyama and Mimura 2010)

experience severe land subsidence are shown in Fig. 6.4. The figure shows the inundated areas obtained by assuming a sea-level rise of 88 cm by the end of the twenty-first century under a A1B scenario, as described in the Special Report on Emissions Scenarios (SRES) (IPCC 2007). Figure 6.3 shows areas in Southeast Asia that are at risk from land subsidence and SLR. The combined effect (named relative SLR in Fig. 6.5) of land subsidence and SLR is expected to increase the relative SLR, as shown in Fig. 6.4, which in turn increases inundation.

Therefore, precise predictions of time-dependent variations of SLR and land subsidence should be estimated, at least from the present to 2100. Settlement vs. elapsed time relations are predicted using the procedure described in Murakami

Fig. 6.5 Definition of
relative sea-level rise
(Murakami and Yasuhara
2011; Yasuhara et al. 2015)

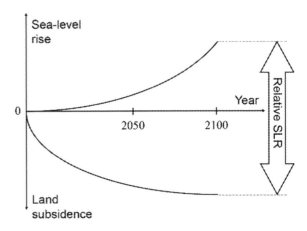

Fig. 6.6 Procedure for
estimating observation-
based predictions of land
subsidence (Murakami
et al. 2006; Murakami and
Yasuhara 2011; Yasuhara
et al. 2015)

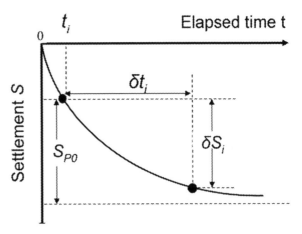

et al. (2006) and shown in Fig. 6.6. Measurements of settlement are required for as
many locations as possible (see Murakami et al. 2006).

The following equation is adopted to estimate the observation-based predictions
of land subsidence (Murakami et al. 2006).

$$\delta S_i = S_{P0}\left\{1 - \exp\left(-C_R \delta t_i\right)\right\} \tag{6.1}$$

where, S_{po} and C_R are parameters statistically determined using measured data at
each location.

6.2.2   Case Study: Mekong Delta, Vietnam

Similar to a case study conducted in the Chao Phraya Delta, Thailand, by Murakami et al. (2006), an attempt was made by Kawase et al. (2012) to predict the effects of sea-level rise and land subsidence in the Mekong Delta in Vietnam, although available data for land subsidence was insufficient and not reliable.

Land subsidence is known to have taken place in the Mekong Delta for many years (e.g, Karlsrud and Vangelsten 2017), but the manner and extent of settlement in the region remain unknown. Here the authors quantitatively describe the present and future situations of land subsidence in the Mekong Delta.

Interferometry Synthetic Aperture Radar (InSAR) was used to investigate land subsidence in the Mekong Delta. Computer software (SIGMA-SAR Cloud Platform) is useful for analyzing the variations in ground surface level caused by land subsidence over time (Murakami et al. 2006). Data for the InSAR dataset were obtained from JERS-1 image data received from December 28, 1993 through January 1, 1997. The data were modified to adapt to the SRTM3 (Shuttle Radar Topography Mission, Ver.3) and were used to produce a digital elevation model.

Land subsidence (LS) was inferred from the InSAR results and representative regions that experienced land subsidence in the Mekong Delta were identified. Finally, future settlement in the representative regions was predicted as described by Kawase et al. (2012).

The land subsidence map produced using InSAR shows that extensive settlement has occurred in Ho Chi Minh City (HCM), My Tho, and Can Tho in the period 1996–1998. The settlement rate was approximately 5–10 mm/year in HCM and around 15–20 mm/year in My Tho and Can Tho. Predictions of land subsidence in HCM, My Tho and Can Tho until 2100 appear to differ, with settlement continuing until 2100 in HMC and settlement being unlikely in My Tho and Can Tho in the future.

LS is accelerated by human activities such as excessive abstraction of groundwater. Therefore, adaptation to observed settlement over time is important as the extent of inundation damage increases due to SLR and LS. Figure 6.7 shows the following inundation scenarios in the Mekong Delta: the present, the future considering SLR, and the future considering both SLR and LS. The results emphasize the importance of considering not only the influence of SLR, but also that of land subsidence in reducing the extent of inundated areas.

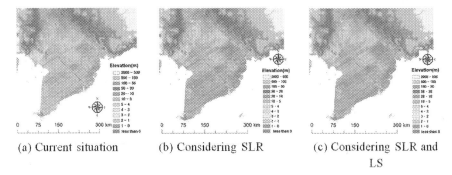

(a) Current situation   (b) Considering SLR   (c) Considering SLR and
                                                                       LS

Fig. 6.7 Predicted inundation in Mekong Delta of Vietnam for the end of twenty-first century (Kawase et al. 2012) (**a**) Current situation (**b**) Considering SLR (**c**) Considering SLR and LS

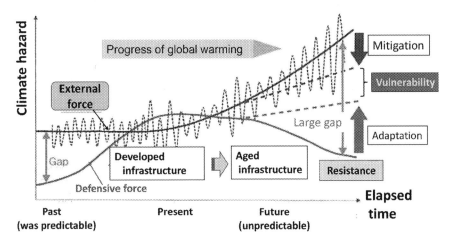

Fig. 6.8 Definition of resilience (modified from Komatsu et al. 2013)

6.3 Resilience and the Increase Thereof

6.3.1 Definition of Resilience

In the current manuscript, resilience is defined as the human potential that is capable of minimizing the gap between the external force caused by climate change and the defensive actions against climatic events, as shown in Fig. 6.8. This gap is sometimes called "vulnerability". Therefore, it can be said that resilience decreases vulnerability.

6.3.2   Scientific Considerations Underlying Burying the Gap Between External and Defensive Forces

In addition to mitigation, adaptation, and combination of mitigation with adaptation, the concept of reactive to proactive responses should also be considered. For example, in a case study on flooding in 2015 in which torrential rainfall at Jyoso City in Ibaraki caused extensive damage to infrastructure and residential areas, the Kinu river dyke height was raised from 4.5 m to 6.0 m. As a result, the following active measures have been proposed by The Ministry of Land, Infrastructure, Transportation and Tourism (MLIT) as mitigation measures against heavy rainfall (see Fig. 6.9).

(i)   Drain work at dyke heels
(ii) Replacement of high-quality soils for avoiding differential settlement
(iii) Installation of impermeable sheet pile walls to prevent river water seepage
(iv) Surfacing work combined with concrete blocks and permeable sheets
(v)   Paved top for preventing rainwater seepage

Generally, the methodology for strengthening river dykes is divided into the following steps: (1) structural reinforcement using pile installation and geosynthetics and (2) improvement of dyke soils mixed with additive materials such as cement and/or fibers (Sato et al. 2013; Yasuhara 2016). Combined adaptation is ideal, but their selection depends on government decisions. The method adopted by the MLIT can be classified as structural reinforcement (Fig. 6.9). However, the authors conducted numerical analysis to propose how dyke soils could be improved to increase overall dyke stability. Results of numerical analyses demonstrate that increased cohesion is the most important contributing factor for increased stability of sandy dykes undergoing water level rise (WLR). From a practical perspective, the results

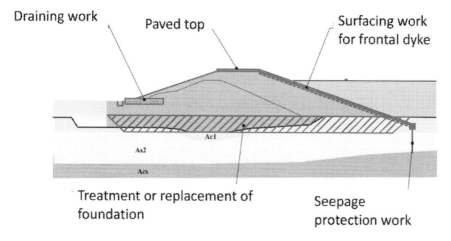

Fig. 6.9 Countermeasures for remediation of damaged river dykes (after MLIT, Japan)

imply that adaptive measures such as adequate compaction of dyke soils at the deep layer and addition of cement to the shallow layer increase the stability of dykes undergoing WLR and overflow or overtopping. Based on the aforementioned flooding in Joso City, the following can be proposed to prepare for future extreme events (Fig. 6.10).

(i) Information collection and dissemination should be increased through cooperation with local community leaders, who are responsible for obtaining the cooperation of local residents.

(ii) Information transmission systems should be established using various tools, such as wireless-activated disaster warning systems, the internet, and the periodic patrols by public relations vehicles.

(iii) It is necessary to activate voluntary organizations for disaster prevention and to promote the adoption of information transmission systems related to flood damage. Response strategies for aged persons, handicapped persons, and foreigners in emergencies should be shared among stakeholders, such as local communities and local governments.

(iv) Early warning and early evacuation awareness need be increased, and residents' understanding of where and how they should evacuate in emergency should be reinforced.

Regarding item (i) above, important information exists in laboratory test results on the vulnerability of river dykes to erosion in Kanto, as shown in Fig. 6.11 (Sato et al. 2013). The results presented in Fig. 6.11 would have worked well for implementation as proactive adaptive measures for unexpected extreme floods if the government had focused on such measures, because they indicated which rivers and which dykes in the Kanto region were vulnerable to erosion.

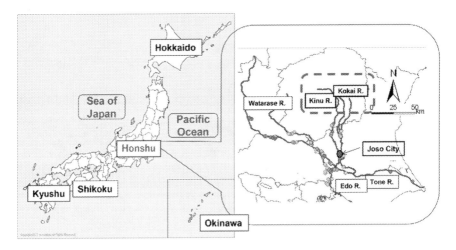

Fig. 6.10 Location of Kinu River and Joso City which suffered flood in 2015. Copyright(C) T-worldatlas All Rights Reserved

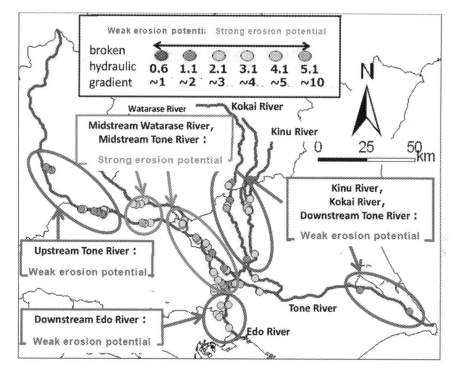

Fig. 6.11 Erosional vulnerability assessment of river dykes in the Kanto region (Sato et al. 2013)

The problem illustrated by Fig. 6.12 is who should play a role in connecting each stakeholder and their respective responsibilities. Although Fig. 6.12 shows that it is the responsibility of mass media, NGOs or NPOs, would it easy to obtain agreement among all stakeholders if a university is charged with the responsibility? This remains uncertain and is open to further discussion.

6.4 Lessons Learned

(i) Inundation of coastal regions undergoing a combination of SLR and LS is more serious than in regions without LS. In such regions, multiple adaptive measures should be adopted.

(ii) The concept of adaptation should be shifted from being reactive to proactive, as doing so would lead to increasing overall resilience.

(iii) Successful adaptation essentially not only requires domestic networks to connect stakeholders, but also the international networks.

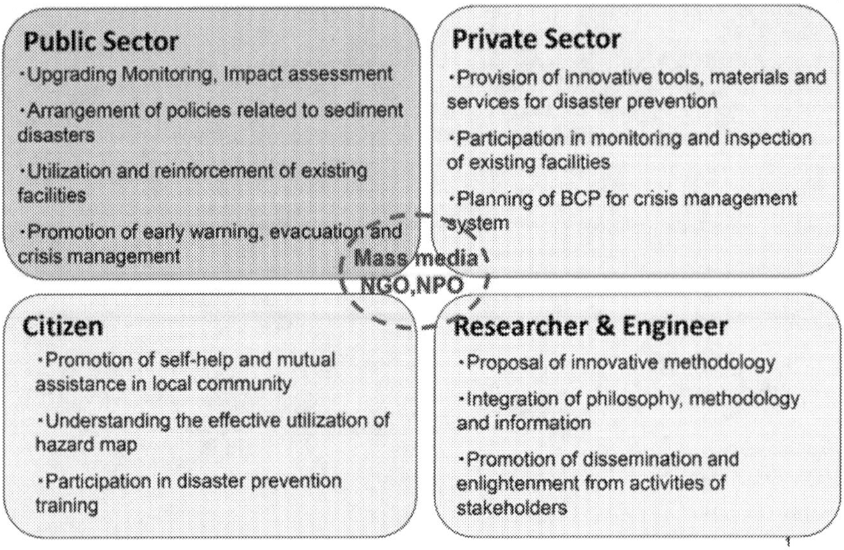

Public Sector
- Upgrading Monitoring, Impact assessment
- Arrangement of policies related to sediment disasters
- Utilization and reinforcement of existing facilities
- Promotion of early warning, evacuation and crisis management

Private Sector
- Provision of innovative tools, materials and services for disaster prevention
- Participation in monitoring and inspection of existing facilities
- Planning of BCP for crisis management system

Mass media
NGO, NPO

Citizen
- Promotion of self-help and mutual assistance in local community
- Understanding the effective utilization of hazard map
- Participation in disaster prevention training

Researcher & Engineer
- Proposal of innovative methodology
- Integration of philosophy, methodology and information
- Promotion of dissemination and enlightenment from activities of stakeholders

Fig. 6.12 Network for connecting stakeholders for disaster damage reduction

Acknowledgements The author expresses his sincere gratitude for financial support from a Grant-in-Aid from MEXT under the supervision of Professor Tetsuji Ito, the previous Director of Institute for Global Change Adaptation Science (ICAS), Ibaraki University, Japan.

References

Intergovernmental Panel of Climate Change (IPCC) (2007) Climate Change 2007: impacts, adaptation, and vulnerability, https://www.ipcc.ch/assessment-report/ar4/

Karlsrud K, Vangelsten BV (2017) Subsidence and land loss in the Ca Mau Province – Vietnam. Causes, consequences and mitigation options. Geotech Eng J SEAGS AGSSEA 48(1):26–32. ISSN 0046-5828

Kawase M, Murakami S, Komine H (2012) Monitoring of land subsidence in the Mekong Delta by using interferometry SAR. In: Proceedings of Awam international conference on civil engineering (AICCE'12) Geohazard information zonation (GIZ'12), Malaysia, pp 790–793

Kokusho T (2005) Extreme events in geohazards in Asia. In: Proceedings of International Conf. on Geotechnical Eng. for Disaster Mitigation & Rehabilitation, Singapore, pp 1–20

Komatsu T, Shirai N, Tanaka M, Harasawa H, Tamura M, Yasuhara K (2013) Adaptation Philosophy and strategy against climate change-induced geo-disasters. In: Proceedings of 10th JGS Symp. on Environmental Geotechnics, Tokyo, Japan, pp 76–82

Maruyama Y, Mimura N (2010) Global assessment of climate change impacts on coastal zones with combined effects of population and economic growth. Selected Papers of Environmental Systems Research, JSCE 38:255–263. (in Japanese)

Mimura N (2013) Sea-level rise caused by climate change and its implications for society – review. Proc Jpn Acad Ser B 89(7):281–301

Murakami S, Yasuhara K (2011) Inundation due to global warming and land subsidence in Chao Phraya Delta. In: Proceedings of the 14th Asian Regional Conference on Soil Mechanics and Geotechnical Engineering, CD-ROM, Hong Kong

Murakami S, Yasuhara K, Suzuki K, Komine H (2006) Reliable land subsidence mapping using a spatial interpolation procedure based on geo-statistics. Soils Found 46(2):123–134

Sato K, Komine H, Murakami S, Yasuhara K (2013) Experimental evaluation of seepage failure of river dykes with natural Fiber mixed with soils. Proceedings of Geotech-Hanoi, Hanoi, Vietnam, pp 1–8

Yasuhara K (2016) Geotechnical responses to natural disasters and environmental impacts in the context of climate change. In: Proceedings of International Conference on Geotechnics for Sustainable Infrastructure Development – Geotech Hanoi 2016, Phung (Editor), Hanoi, Vietnam, pp 957–981

Yasuhara K, Murakami S, Mimura N (2015) Inundation caused by sea-level rise combined with land subsidence. Geotech Eng J SEAGS AGSSEA 46(4). December 2015, ISSN 0046-5828

Chapter 7
Climate Change Adaptation in Fisheries Livelihoods Associated with Mangrove Forests in Xuan Thuy National Park, Vietnam: *A Case Study in Giao An Commune, Giao Thuy District, Nam Dinh Province*

Thu Hoai Nguyen

7.1 Introduction

Giao An, a commune located in the southeast of Giao Thuy District, Nam Dinh Province in Vietnam, is one of five communes in the buffer zone of Xuan Thuy National Park with a history associated with the development of the Red River Delta. About 300 to 400 years ago, this land was comprised of pristine beaches with lots of reeds and mudflats. Over time, the people of Giao An built dikes, encroached on the sea, and leveled the land. By the end of 1860, Giao An was established, comprising Trung Uyen, Hanh Thien, Xuan Hy, Thuy Nhai, Hoanh Lo, Tra Huong, and Tra Lu villages (divided into 16 hamlets). The period from 1960 to 1985 was characterized by continued expansion of arable land by building dikes under the motto, "Rice encroached on sedge, sedge encroached on *Bruguiera* (genus of mangrove plants), *Bruguiera* encroached on the sea". At this stage, the commune had reclaimed approximately 300 ha of land from the sea, close to the foot of Ngu Han dike, and established six hamlets in the Dien Bien New Economic Zone Area (from hamlet 16 to hamlet 22). The period from 1985 to 1995 was characterized by the adoption of the marine economic development strategy under the motto "*Bruguiera* encroached on the sea, shrimp encroached on *Bruguiera*", which created thousands of hectares of shrimp ponds in Bai Trong and Con Ngan. Due to economic development, the natural mangrove forests in Giao An Commune decreased, especially in Bai Trong.

However, realizing the important role of mangroves in coastal area protection, the local government has implemented numerous solutions, simultaneously, such as planting more than 1000 ha of mangrove in Con Ngan and a part of Con Lu from 1997 to 2003. In addition, Giao An has pioneered the development of climate

T. H. Nguyen (✉)
VNU Vietnam Japan University, Hanoi, Vietnam

© The Author(s) 2022
T. Ito et al. (eds.), *Interlocal Adaptations to Climate Change in East and Southeast Asia*, SpringerBriefs in Climate Studies,
https://doi.org/10.1007/978-3-030-81207-2_7

change (CC) adaptive fisheries livelihood models, taking advantage of mangroves which are improved extensive shrimp farming and community-based mangrove management. These models have not only improved resilience and response to CC, but they have also stabilized the incomes of local people as well as developed the area by making economic activities more sustainable.

The study used the following three approaches to investigate CC adaptation in fisheries livelihoods associated with mangrove forests in Giao An Commune: (1) systematic, interdisciplinary approach; (2) combined top-down and bottom-up approach; and (3) Department for International Development's sustainable livelihoods approach. Furthermore, tools from sociological research methods (e.g., questionnaires for households) and Participatory Rural Appraisal (in-depth interviews as well as establishment of seasonal calendars and a calendar of events) were used. One hundred households and twenty-four civil services were interviewed to collect relevant information.

7.2   Results and Discussion

The average annual temperature fluctuated between 23 °C and 24 °C. In 30 years, it tended to increase by about 1.0 °C (Fig. 7.1). In particular, from 1989 to 1998, the average annual temperature increased by 0.4 °C, while from 1999 to 2008, the temperature decreased by 0.2 °C, and the remaining period from 2009 to 2018 it increased by 0.6 °C per annum.

Moreover, the total annual rainfall of Nam Dinh Province from 1989 to 2018 fluctuated in the range of 1750–1800 mm. 2007 was the year with the lowest annual rainfall (1086 mm) and 1994 was the year with the highest total annual rainfall (2988 mm) (Fig. 7.2).

All surveyed households confirmed that the natural disasters in Giao An Commune in recent years have been atypical, with "typhoon"; "tropical depression"; "lightning"; "heavy rainfall"; "extreme hot weather"; "extreme and damaging cold" and "hoarfrost" being considered to be the main natural disasters in the study area (Table 7.1). The frequency of most of these above natural disasters was rated as "rare", but the intensity was "stronger" than before.

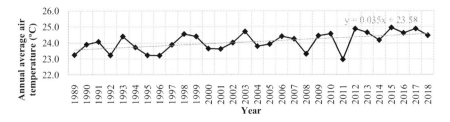

Fig. 7.1 Trend in average annual air temperature in Nam Dinh Province in the period 1989–2018. Source: Department of Natural resources and Environment of Nam Dinh Province (2019)

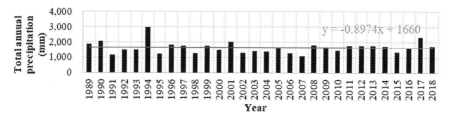

Fig. 7.2 Trend in total annual rainfall in Nam Dinh Province in the period 1989–2018. Source: Department of Natural resources and Environment of Nam Dinh Province (2019)

Table 7.1 Assessment of survey respondents about natural disasters that have occurred in Giao An Commune in recent years

Types of natural disasters	Valid percentage (%)	Types of natural disasters	Valid percentage (%)
Typhoons	97.0	Hail	4.0
Tropical depressions	98.0	Drought	30.0
Whirlwinds	56.0	Hoarfrost	82.0
Lighting	94.0	Other	20.0
Heavy rainfall	78.0	- acid rain	1.0
Inundation	32.0	- showers	1.0
Sea level rise	19.0	- fog	1.0
Saltwater intrusion	19.0	- change in weather	16.0
Extreme hot weather	95.0	- humidity	1.0
Extreme and damaging cold	60.0		

During the survey, in addition to in-depth interviews with local people and authorities, the impacts of natural disasters were reported as follows (Table 7.2).

In addition, agriculture and fisheries were the main livelihood activities accounting for a large proportion of the local economy (67%) (People's Committee of Giao An Commune 2017). In terms of fishery production, two activities are very important: aquaculture and capture fisheries (or called community-based mangrove management). The entire commune utilized an aquaculture area of 669.2 ha, which comprised a freshwater area of 59 ha and a shrimp pond area of 610.2 ha. The total aquatic production in 2019 was estimated at 2200 tons (People's Committee of Giao An Commune 2019).

a. Aquaculture

Peaks in aquaculture production coincided with the different types of natural disasters. Consequently, without careful preparation, the risk to aquaculture households was quite large. The natural disasters markedly affected these activities in the Giao An Commune, especially "typhoon", "tropical depression", "heavy rainfall", "saltwater intrusion", "extreme hot weather", "extreme and damaging cold", "hoarfrost" and "other", which comprised changes in climate and weather. Together, these phenomena "reduced productivity", "changed the water environment", which can lead to "lost all aquaculture products". Table 7.3 shows the valid percent of

Table 7.2 Calendar of natural disasters in Giao An Commune

Year	Types of natural disasters	Damage
2010	Extreme and damaging cold lasting 38 days	100% of the areas of vegetables, rice seeds, and rice were damaged; many aquaculture households were affected.
2011	Extreme and damaging cold	Newly transplanted rice fields died or were poorly developed.
2012	Typhoon son Tinh (typhoon no. 8)	- 60% of pure rice which had not been harvested was damaged. - 100% of vegetable areas were damaged - 50% of clam farming was damaged. - 100% of auxiliary works were damaged. - more than 80% of freshwater aquaculture areas were damaged. - the roofs of 100% of corrugated iron houses were destroyed. - 3 houses collapsed.
2013	Typhoon no. 1, 3, and 6	Vegetable and rice areas were affected.
	Extreme hot weather	Shrimps grew slowly due to reduced natural food intake and hot weather; clams experienced severe heat stress and died on hot days.
2014	Heavy rainfall	Salt concentration in aquaculture areas fell below 10.0‰ leading to shrimps and clams dying and growth retardation.
2015	Northeast monsoon and damaging cold	Impacted on production activities, with aquaculture and fishing households severely affected.
2016	Extreme and damaging cold	Affected production activities.
	Typhoon no. 1	- 100% of vegetable and transplanted rice field, as well as aquaculture ponds were damaged. - many cattle and poultry died. Total losses amounted to 26.7 billion VND.
2017	Typhoon no. 2, 4 and 10	Tropical depression caused heavy rain in combination with the water from Hoa Binh hydropower reservoir to create floods in the river, which adversely affected agricultural production in Giao an.
2018	Typhoon no. 4 combined with lightning	- adversely affected production activities and daily life of the people. - some household TVs were broken by lightning.
2019	Typhoon	Affected production activities and daily life of the people.

impacts of natural disasters on aquaculture households that are engaged in fish, shrimp, and clam farming.

b. Capture fisheries

Capture fisheries were exploited regularly depending on the health of the participants, weather conditions as well as the tidal calendar. However, surveyed households emphasized that weather and climate were reasons for the decline in aquatic capture production. Not only that, if "typhoon", "tropical depression", "heavy rainfall", "extreme hot weather" or "extreme and damaging cold" occurred, households would temporarily also stop production activities (Tables 7.4 and 7.5).

Due to the above impacts and the complexity of CC, Giao An deployed several CC adaptive fisheries livelihood models taking advantage of mangroves. These

Table 7.3 Assessment of the effects of natural disasters and CC impacts on aquaculture

							(Units: %)
Types of natural disasters	Slow growth	Reduced productivity	Changes in water environment	Numerous diseases	Difficulty eating	Lost all aquaculture products	Dead aquatic species
Typhoons	3.6	39.3	7.2	0.0	0.0	14.3	0.0
Tropical depression	0.0	17.9	0.0	0.0	0.0	7.2	0.0
Heavy rainfall	0.0	32.1	32.1	0.0	0.0	0.0	0.0
Sea level rise	0.0	0.0	39.3	0.0	0.0	21.4	0.0
Extreme hot weather	10.7	0.0	0.0	10.7	10.7	7.1	42.9
Extreme and damaging cold	0.0	0.0	0.0	0.0	0.0	39.3	32.1
Hoarfrost	0.0	0.0	0.0	0.0	0.0	39.3	14.3
Other	0.0	0.0	0.0	0.0	0.0	7.2	3.6

models improved extensive shrimp farming and community-based mangrove management.

c. Improved extensive shrimp farming

Since the 1980s, the wetlands of Giao Thuy District (formerly Xuan Thuy district) had been converted into fishing and shrimp ponds. Giao An was a leading commune in this movement. By the mid-1990s, due to declining natural aquatic resources, people began to develop improved extensive shrimp farming. This model started to become more widely implemented at this time.

The basic technique of the model is described as follows: Black tiger shrimp are stocked in small ponds that are built inside the main ponds where they are fed a certain amount of industrial feed. Then, when the weight reaches 35,000 shrimp/kg, or after about 1 week, feeding is stopped and only food from nature is used. In the process of raising shrimp, farmers still can exploit other aquatic species from the wild. Typically, the value obtained from natural resources accounts for about 50% of the total income. When implementing this model, it is necessary to ensure that the mangrove forests cover an average of 30% of the total aquaculture area.

The improved extensive shrimp farming model associated with mangrove forests in Giao An Commune was assessed to "increase resilience to weather and CC" (87%); "improve life for local people" (72%) and "stabilize income of participating households" (71%) (Fig. 7.3). Compared with the traditional extensive shrimp farming model (without adding food during aquaculture), this model has higher productivity and economic efficiency. For the current intensive whiteleg shrimp

Table 7.4 Seasonal calendar in fisheries activities of households in Giao An Commune (according to the lunar calendar)

	Jan	Feb	Mar	Apr	May	Jun	Jul	Aug	Sep	Oct	Nov	Dec
I. Fisheries activities												
1. Capture fisheries	x	x	x	x	x	x	x	x	x	x	x	x
2. Extensive shrimp farming[a]	x Renovate pond	x First stock	x Second stock (if any)	x	x	x First harvest	x First harvest	x	x Harvest all	x Renovate pond	x Renovate pond	x Renovate pond
II. Natural disasters												
1. Typhoon, tropical depression						x	x	x	x	x		
2. Heavy rainfall						x	x	x	x			
3. Extreme hot weather					x	x	x					
4. Hoarfrost									x	x	x	
5. Extreme and damaging cold	x										x	x

[a] Depending on the households, shrimp can be stocked once or several times

Table 7.5 Summary of impacts of natural disasters due to CC on the fisheries livelihoods of Giao An Commune

Types of natural disasters	Impacts on fishery
Extreme hot weather	- the temperature of water supply for aquaculture will increase, which causes aquatic species to die; - Worker's health is affected (by heatstroke).
Heavy rainfall concentrated during the rainy season (inundation)	- dikes and embankments of aquaculture ponds are destroyed or degraded; - the water environment is changed; - productivity reduced; - production loss in aquaculture.
Extreme and damaging cold	The temperature of water supply for aquaculture is cold, which causes aquatic species to die.
Hoarfrost	- productivity reduced; - production loss in aquaculture.

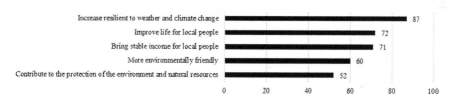

Fig. 7.3 Effectiveness of implementing improved extensive shrimp farming model (%)

farming, the improved extensive shrimp farming has a higher ability to adapt to CC as well as being less harmful to the environment, despite the lower profit (Table 7.6).

d. Community-based mangrove management

Giao An Commune is a pilot locality for a community-based mangrove management model being funded by the Centre for Marinelife Conservation and Community Development (MCD). Up to now, there have been 27 individuals/households participating in this model with a total management area of 567.2 ha that was divided into 14 plots. The participants can harvest some aquatic species under the mangroves, such as mollusks and fish, but must not affect the ecological balance of the area. At the same time, they must strictly comply with regulations on environmental resources protection in the buffer zone of Xuan Thuy National Park, such as not cutting trees or causing forest fires; not changing the landscape; not polluting the environment, not exploiting, destroying, and depleting aquatic resources; not hunting, trapping or chasing other wild animals and birds.

Compared with similar models being implemented in other areas, the new feature of these models is that in Giao An Commune there are linkages for mutual development in production activities. The improved extensive shrimp farming households in Giao An have links with traders. The linkage method is as follows: households record their shrimp production before traders transfer all of those goods to Hanoi. After the calculations, traders will reimburse households (Fig. 7.4).

Table 7.6 Comparing improved extensive shrimp farming model and industrial shrimp farming model (intensive whiteleg shrimp farming)

	Improved extensive shrimp farming model	Industrial shrimp farming model (intensive whiteleg shrimp farming)
Input cost	Lower From 50–150 million VND/pond/year	Higher About 200 million VND/pond/year
Profit	Medium	Large
Payback period	Slow, unfocused	Quick, one time
Stocking density	Lower	Higher and thicker
Type of land used	Tidal mudflat with mangrove	Tidal mudflat without mangrove
Disease	Little, easy to handle	Many, difficult to handle
Food for shrimp	Most food is from nature	All feed is industrial
Water sources	Supplied by nature - tidal resources	Supplied by human
Waste sources	Little Does not affect the environment too much	Requires water treatment
Labor	Part-time	Full time
Adaptation to CC	Higher because of having mangroves	Lower
Long-term future	Sustainable and effective (poly-culture)	Unsustainable (mono-culture) After 2 or 3 crops, must be suspended to limit disease

a. A stocking pond inside improved extensive shrimp farming pond.   *b. Mangrove trees in improved extensive shrimp farming area.*

Fig. 7.4 Some pictures related to improved extensive shrimp farming in Giao An Commune

In capture fisheries, there is a self-governing team. The team was established in 2015 under the leadership of Mr. Pham Van Bong (head of the hamlet 18). As of 2019, 17 ships and boats were operating near the shore participating in this group, including some of the ships in Giao Thien Commune. Each month, the team holds a meeting. The purpose of the team is to support each other with production, and to perform rescues if the boats experience difficulties at sea. In addition, the team is entrusted by the Border Guard to manage the mangrove forests. Every day, in addition to fishing, the team members will patrol the area to check for acts of harming the environment. From there, they notify the forest rangers, the local police, border posts, and local authorities to handle every situation. Moreover, the group also plays a role in loan clubs, with each member contributing two million VND/month. Borrowers must pay the initial amount plus the interest rate of 0.5%/month. This fund is mainly to support the team members if they need to repair ships or equip new fishing grounds.

Community-based mangrove management model not only "protects and restores mangrove forests" (82%); "restricts unsustainable exploitation of forest resources" (80%) but also "creates livelihoods for local people" (77%) and "stabilizes income for participants" (75%) (Fig. 7.5). According to the survey, the average income from this activity ranges from 200–300 thousand VND/day/person.

In addition to the positive effects, these models still have some problems related to human capital and social capital.

e. Human capital

The education level of the survey participants is low, with most people having attended secondary school being the largest group (73%) (Table 7.7). This is a major obstacle in improving the level of human capital and diversifying different types of livelihoods as well as applying science and technology to production.

Regarding training, 6% of the people were selected by the commune and were sent to training, conferences, or programs related to fisheries activities organized by the Commune Agricultural Production Steering Board/Giao Thuy Division of Agriculture and Rural Development/Department of Agriculture and Rural Development of Nam Dinh Province or Red Cross Association and Xuan Thuy National Park. The form of training was focused-training. In particular, more than half of the surveyed people wanted more knowledge, and to gain experience and learn techniques related to aquaculture.

f. Social capital

Although Giao An Commune has basic links to fishery production activities, there is not yet a close linkage or value chain in this sector. For example, in capture fisheries, only a small percentage of households participate in a Self-governing team (17 ships). Or households engaging in extensive shrimp farming have not been

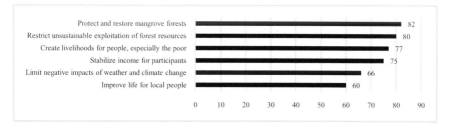

Fig. 7.5 Effectiveness of implementing community-based mangrove management model (%)

Table 7.7 The educational level of survey respondents

Education level		Percentage (%)
Never been to school		2.0
Primary school	Have not completed	2.0
	Completed	4.0
Secondary school	Have not completed	55.0
	Completed	18.0
High school	Have not completed	5.0
	Completed	6.0
Primary vocational school		1.0
College	Secondary apprentice school	1.0
	College	2.0
University		3.0
Academy		1.0

Table 7.8 Surveyed participants assessed the roles of mangroves

Role	Valid percentage (%)	Role	Valid percentage (%)
Disaster prevention in cases of typhoons	87.4	Shelter for aquatic species	12.6
Stabilizing mudflats	51.7	Regulating air	9.2
Controlling rise of sea level and sea currents	17.2	Preventing soil erosion and regenerating soil	3.4
Protection of sea dykes	73.6	Limiting flooding	1.1
Protection of shrimp and clam ponds	44.8	Reducing environmental pollution, protecting the environment	10.3
Retaining sediment and creating mudflats	18.4	Developing ecotourism	3.4
Increasing income for fishermen	51.7		

proactive in setting prices because it is highly dependent on traders and market prices. Therefore, they also cannot avoid facing price squeezes.

7.3   Lessons Learned

Mangroves played an important role in the households of the Giao An Commune. In particular, the roles of "disaster prevention in cases of typhoons" (87.4%); "stabilizing mudflats" (51.7%); "increasing income for fishermen" (51.7%), and "protection of shrimp and clam ponds" (44.8%) all scored highly (Table 7.8).

Giao An has made effective use of mangrove forests to build livelihood development models which have improved extensive shrimp farming and community-based mangrove management. The new issue, compared to other localities, is the linkages in production to promote livelihood development. These models not only bring economic value, but they also contribute to the resilience of the local people to CC as well as ensuring sustainable development in the region.

The models of fisheries livelihoods and how they can be used to adapt to CC associated with mangroves have generated positive results. To continue to develop livelihood activities as well as the models, the local authorities of Giao An Commune need to focus on two livelihood assets; human capital and social capital. For human capital: (1) Organizing training courses, short-term and long-term courses depending on the characteristics of the livelihoods to equip local people with the necessary knowledge and skills in production activities in the context of CC; (2) Organizing community events combined with contests related to CC. For social capital: Studying how to create more effective production groups, teams, and areas. For example, establishing an aquaculture farming group with the task of equipping necessary knowledge and skills in aquaculture through exchanging, sharing experiences as well as supporting and helping each other in areas of finance, mechanization, or equipment.

References

Department of Natural resources and Environment of Nam Dinh Province (2019) Climate assessment report in Nam Dinh Province
People's Committee of Giao An Commune (2017) Report on results of the implementation of new rural construction criteria for Giao An Commune, Giao Thuy District, Nam Dinh Province
People's Committee of Giao An Commune (2019) Report on the implementation of socio-economic development and national defense tasks in the first six months and key tasks for the last 6 months of 2019

Chapter 8
Community-Based Disaster Risk Reduction Education in Japan

Aiko Sakurai and Tetsuji Ito

8.1 Introduction

Community-based disaster risk reduction (CBDRR) is considered to be the core of any risk reduction approach. CBDRR is defined as an approach that seeks to: (1) reduce the vulnerabilities and increase the capacities of vulnerable groups and communities to cope with, prevent, or minimize loss and damage to life, property, and the environment, (2) minimize human suffering, and (3) hasten recovery (Shaw 2016).

CBDRR has been emphasized from the beginning of the United Nations International Decade of Natural Disaster Reduction (1990–1999). From the 1994 Yokohama Plan of Action to the latest Sendai Framework for Disaster Risk Reduction (SFDRR) 2015–2030, a massive change has occurred in the concepts, approaches, and methods used to reduce the impacts of disasters. In the 1990s, the focus was more on multi-stakeholder and local governance; however, since 2000, more risk-sensitive investment planning has been emphasized, and risk-informed decision-making has become the core of risk reduction (Shaw 2016). In relation to understanding risk, the SFDRR emphasizes that disaster risks have local and specific characteristics that must be understood to devise measures that reduce disaster risk. Since the scale and degree of a natural disaster depend on local conditions, each community needs to enhance its capacities (UNISDR 2015).

A. Sakurai (✉)
Department of Social Sciences, Toyo Eiwa University, Kanagawa, Japan

International Research Institute of Disaster Science, Tohoku University, Miyagi, Japan
e-mail: sakurai.aiko@toyoeiwa.ac.jp

T. Ito
College of Humanities and Social Sciences/Global and Local Environment
Co-creation Institute, Ibaraki University, Ibaraki, Japan

© The Author(s) 2022
T. Ito et al. (eds.), *Interlocal Adaptations to Climate Change in East and Southeast Asia*, SpringerBriefs in Climate Studies,
https://doi.org/10.1007/978-3-030-81207-2_8

In Japan, there are three types of help for disaster risk reduction (DRR): self-help, mutual help, and public help. Owing to the massive scale of the 2011 Great East Japan Earthquake (GEJE) and tsunami disaster, municipal governments were severely affected, and their ability to help people was paralyzed. In this situation, community residents were responsible for supporting each other's survival, such as through evacuating from tsunami-affected areas, operating evacuation shelters, and securing adequate supplies of food and water. In essence, communities became first responders. Due to the recognition of the limitations of public help after this calamity, the importance of self-help and cooperation within each community has been reemphasized. Since the GEJE, accurate risk assessment by the government and experts, as well as interactive risk communication among the government, experts, and community residents, has been reemphasized to gain a better understanding of local disaster risks and take appropriate action in times of emergency. Based on the 2011 disaster experience and the lessons learned from it, how to protect individual lives effectively and respond practically to catastrophic situations at the local level are key issues for enhancing community resilience (Sato et al. 2018).

Acknowledging such a trend in international strategies and national level policy arrangements in Japan, this chapter examines three Japanese case studies on CBDRR and DRR education—in Sendai and Ishinomaki Cities in Miyagi Prefecture and in Joso City in Ibaraki Prefecture (Fig. 8.1)—to extract some lessons learned that could apply to community-level adaptations to climate change in East and Southeast Asian countries.

Fig. 8.1 Location map of the three case studies

8.2 Development of Community-Based Human Resources for Disaster Risk Reduction (DRR)

8.2.1 Promoting Sustainable Community-Based Disaster Risk Reduction (CBDRR) in Community Development

In Japan, mutual help has been promoted through activities by community-based disaster management organizations (*jishu-bousai-soshiki* or *jishubo*), which are volunteer-based community governance units within neighborhood associations (*chonaikai*). Local government agencies make full use of these organizations to transmit information and instructions to neighborhood residents. Although neighborhood associations in Japan have long histories and deep roots in their communities, their capacities have been weakened, and they increasing face the prospect of disappearing due to urbanization, the diversification of lifestyles, and the aging of the population. Therefore, although the *jishubo* to household ratio remains high, individuals are not necessarily motivated to volunteer, and the role of the *chonaikai* leader rotates among its members. Moreover, members have passive attitudes and lack a proactive approach toward promoting disaster preparedness in the community (Kuroda 1998). In this situation, DRR efforts represent an additional burden imposed on these neighborhood associations, on top of their regular daily activities, including neighborhood cleaning, garbage collection, and recycling, as well as organizing community activities such as athletic and cultural events (Sakurai and Sato 2017).

The case of Katahira district, Sendai City, Miyagi Prefecture is an example of an innovative approach to develop sustainable CBDRR. Katahira district has a long history of development with the castle town of Sendai Han (a feudal domain). The southwest side of the district is bordered by the meandering Hirose River, while the east side connects to the central business district of Sendai City, including Japan Rail Sendai Station. The major local disaster risks include earthquakes and flooding. The main body to promote DRR is the Katahira Community Development Association (KCDA), which was formed in 2013 as an organization to implement community development plans representing four major goals: establishing safety and security, energizing the community, conserving and making active use of history and the environment, and developing a sustainable community development system. Within the KCDA, the Katahira Union Neighborhood Association, a union of nine *chonaikai* in the district, plays a central role in aligning various related organizations.

The KCDA responded to the GEJE disaster by establishing and managing evacuation facilities in their district. Katahira-cho Elementary School was designated as an evacuation center by the city and opened as a shelter to accommodate about 1500 evacuees. The original capacity of the shelter, which was mainly for local residents, was 350 evacuees; however, 85% of all the evacuees, including tourists, business travelers, employees at neighboring businesses, and foreign students, were stranded. Having such experiences, since 2011, the KCDA has been working increasingly harder on strengthening their DRR capacities by developing disaster response, disaster survival, and shelter management manuals, conducting emergency drills, and transmitting DRR information to local residents, including foreigners, in cooperation with related local stakeholders (Sato et al. 2018). These DRR activities were led by community disaster management leaders, called *Sendaishi-chiiki bosai leaders* (SBLs), who are trained and certified by the Sendai City government to serve their community's DRR activities. In addition to adult activities, school-aged children from elementary to high school formed a Katahira district community development team for children in 2015 under the KCDA to conduct community development activities.

As a model project under the fiscal year 2016 Cabinet Office Community Disaster Management Plan, the KCDA developed a CBDRR education activity that involved the entire community, called the "Disaster Risk Reduction X Treasure Hunting Game (DRRTHG)", to foster human resources from the next generation. As a result, treasure hunting conducted by event planning firms and local municipalities to revitalize the community and attract tourists became a popular activity in Japan. In a basic procedure of the treasure hunting game, the participants follow a "treasure map" to explore the town and search on foot for a treasure chest hidden at a location in the town. The DRRTHG was developed by the KCDA as an application of a treasure hunting game for DRR. In the DRRTHG, the SBLs and KCDA members guide elementary and junior high school children across the district to explain the abundant nature, culture, and resources for DRR by walking around in the manner of a treasure hunting game. The "Crescent Corps" certificate, on which five rules to become a good community member are printed, is awarded to the children who participate in this activity and pass the evaluation test; these children are expected to become the local human resources for next-generation DRR. The "Crescent Corps" is named after the crescent that was decorated on the front side of *Masamune Date*, a founder of Sendai Han. These rules are fundamental to the concepts of self-help and mutual help, and to the spirit of incorporating DRR into community development planning.

The KCDA continued to implement the DRRTHG from 2016 to 2019 by changing the search area in the district. In the 2019 implementation, elementary school children who participated in the first DRRTHG became senior high school students and served as guides to the participants from local elementary and junior high schools. Fig. 8.2 shows the Katahira DRR survival map (left) and a photo of participants in the 2016 DRRTHG event (Sato et al. 2020).

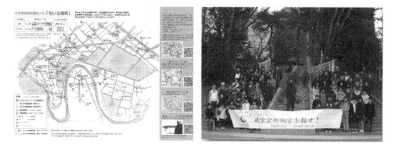

Fig. 8.2   Examples of Katahira District's disaster risk reduction (DRR) activities

8.3   Understanding Local Disaster Risk by Utilizing Geographical Maps at School

The second case is a school-based DRR education program developed in Ishinomaki City, Miyagi Prefecture, one of the coastal municipalities most severely affected by the GEJE and tsunami. Ishinomaki City experienced the tragic Okawa Elementary School Incident, in which 74 of 108 pupils under school supervision, 10 teachers, and 175 local residents were lost as a result of the tsunami that came upstream along the Kitakami River. The Okawa Elementary School Incident resulted in a lawsuit in which the Sendai High Court ruled that the school had failed to meet their obligations to designate a third tsunami evacuation area and to clarify evacuation areas and routes in its risk management manual in advance, and was therefore guilty of negligence. This incident taught important lessons, such as that each school should foresee its own local disaster risk and that all schools should prepare for effective and practical school disaster safety based on the school district's risk assessment with the support of the municipal government and DRR experts (Sakurai 2021).

The Reconstruction and DRR Mapping Program (R-DRRMP) is a school-based DRR education program developed in Ishinomaki in 2012. The R-DRRMP consists of orientation, town-watching, map-making, and presentation of the produced maps. For the first 3 years of the implementation, it was aimed at recording the progress of reconstruction at one of the tsunami-affected schools in the coastal area. A survey was developed with the ninth-grade students at a junior high school in 2018 to follow up the fourth-grade pupils who conducted the Mapping Program in 2013. The results confirmed that they had more pride and affection and were more motivated to think about the future of their community's development than were students from the same grade who had not participated in the R-DRRMP (Sakurai et al. 2020).

A teachers' guidebook on the R-DRRMP published in 2016 helped schoolteachers localize the contents according to the geographical and socioeconomic characteristics of the school district (Tohoku University, 2018). This guidebook contains a teaching plan for incorporating the utilization of maps into town-watching and map-making activities. Based on the guidebook, a teacher training program on how to understand local disaster risks at school by utilizing topographical maps was started

 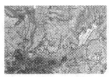  

Fig. 8.3 Sample maps at a school district in Ishinomaki City
(Source: [Left] edited by authors based on a topographic classification map created by the Geospatial Information Authority of Japan; [Center left] edited by authors based on the Kitakami River inundation forecast map created by the Ministry of Land, Infrastructure, Transport and Tourism; [Center right] edited by authors based on a landslide hazard map created by the Ishinomaki City government; [Right] photo of a disaster risk reduction (DRR) map created by a junior high school in Ishinomaki City)

in 2019 (Sakurai et al. 2019). This training program aims at enhancing map-reading skills and fostering teachers' capacity to understand local disaster risks, even beyond those expected, through the reading of topographical and hazard maps. Teachers who participate in the training are expected to utilize these skills in their school's DRR education and improve their school disaster response manual.

Figure 8.3 shows a series of maps prepared for a school district located inland of Ishinomaki City nearby the Kitakami River. The far right panel shows a DRR map made by seventh-grade students in 2019. Students compared these maps in the classroom and conducted town-watching by interviewing local residents to determine the actual damage caused by Typhoon Hagibis in their school district in 2019. The red-colored area indicates the flooded area. The R-DRRMP in Ishinomaki City thus became one of the main approaches for teachers and students to gain a better understanding of their local disaster risks.

8.4 Local Social Networking Service (SNS) Broadcasting System for Facilitating Communication Among Local Residents and a Customized Evacuation Action Plan ("My-TimeLine")

In September 2015, heavy rains in the Kanto and Tohoku regions caused massive flooding in Joso City, Ibaraki Prefecture. Vast fields occupy the area between the Kinugawa and Kokaigawa Rivers, and Mt. Tsukuba (877 m above sea level) can be seen in the distance. Ground level is lower than the river level in some places. Historically, flooding has occurred in this area many times. When the Kinugawa embankment broke during the rainfall of 2015, one-third of Joso City was inundated—even the first floor of the relatively new city hall was flooded. In addition, more than 5000 houses were completely or partially destroyed and more than 4200 people needed to be rescued.

Many residents of Neshinden district in Joso City were able to take refuge before the flooding and earlier than the residents in other areas. One of the reasons for the earlier evacuations in this district was the utilization of a local social networking service (SNS) broadcasting system, "Neshinden Hot Mail", which has been in operation in the district since 2014. Before 2014, the leader of the *chonaikai* had to call each household individually (Wagamachi Neshinden (My hometown Neshinden) (n.d.)). The introduction of the "Neshinden Hot Mail" system allowed the local *chonaikai* leader to deliver short textual information to all local residents at the same time. The new system makes it possible to display textual information on mobile and smartphones through a single operation on a personal computer.

As the rainfall was becoming increasingly heavy, the local leader sent a series of short text messages informing residents of the Neshinden district about the flood stage of the Kinugawa River and issuing evacuation warnings. These SNS messages urged local residents to take early evacuation actions. The success of this system highlighted two main points:

1. Receiving information is possible with only a mobile phone

   Even though the SNS broadcasting is only one-way communication, local residents, including older people, could receive information easily and instantly on their phones. As most of the older people in the area have simple mobile phones as opposed to high-performance smartphones, all they needed to do to be able to receive messages was register their phone number.

2. SNS is routinely used as a means of communication in the community

   SNS has been used as a daily communication channel to share information about district events and condolences, so residents have become accustomed to using it. Therefore, it was a reliable communication tool during the emergency.

Although this idea is very simple, the "Nishiden Hot Mail" example shows the potential of SNS to become a powerful communication tool to facilitate CBDRR. As discussed earlier in the chapter, in Japan, the local community has been declining, and thus, the capacities of the *chonaikai* have been weakened. In addition, people are increasingly likely to not know their neighbors' names or faces. The SNS broadcasting system does not allow two-way communication, but in the areas where there are still trusted local leaders, even one-way communication could help facilitate communication among the *chonaikai* networks. The SNS broadcasting system could have worked to disseminate risk and evacuation information during the emergency in Neshinden district to encourage "mutual help" in regard to CBDRR.

However, to introduce such a system, the following steps are required: (1) have a trusted local leader; (2) obtain the understanding of the residents to introduce it; and (3) be prepared for some required costs (e.g., contracts with a company that has a system for sending Hotmail).

In addition, in Joso City, another advanced example of CBDRR, called "My-TimeLine" (Fig. 8.4), was developed by the River Office of the Japanese Ministry of Land, Infrastructure, Transport and Tourism based on lessons learned from the 2015 disaster. "My-TimeLine" formulates evacuation action plans from

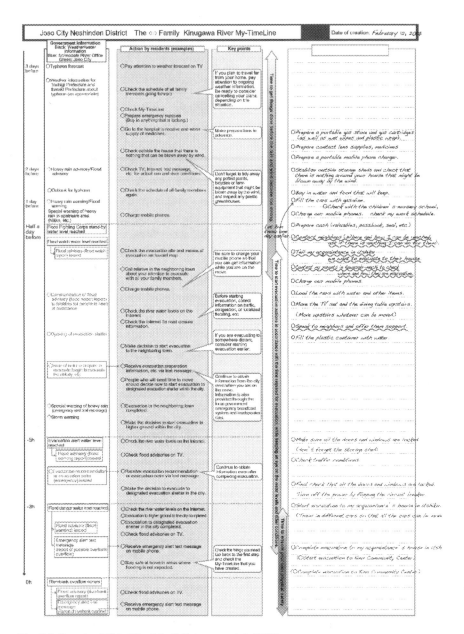

Fig. 8.4 Sample of "My-TimeLine" (Satomura et al. 2020)

3 days before a typhoon or when heavy rain is expected based on the local disaster risk of each household. When a typhoon is approaching, weather and evacuation information are released by the Japan Meteorological Agency and local governments, and people are expected to take appropriate action. However, even if people

receive evacuation information, they may be unsure about the timing of evacuations, how to avoid risks during evacuation, and who to evacuate with. For example, if a person with a disability is in the family, earlies evacuation actions will be required, and if a family member has a chronic disease, it will be necessary to secure medicine. The "My" in "My-TimeLine" implies that the information should be "customized for individual circumstances".

The "My-TimeLine" tool was developed following the flood in Joso City. A version for children was also created, and this is used in disaster reduction classrooms in the area and disaster reduction education in schools. Piloted in Joso City, "My-TimeLine" now enjoys widespread use in Japan, where the risk of flood damage caused by powerful typhoons is increasing because of climate change. For more details on "My-TimeLine", please refer to Satomura et al. (2020).

The combination of the SNS broadcasting system and "My-TimeLine" in Joso City could greatly contribute to enhancing CBDRR in Japan.

8.5 Lessons Learned

As discussed in the beginning of this chapter, CBDRR is thought to be the core of any risk reduction approach, and local disaster risks must be clearly understood to determine community countermeasures. In Japan, the expectation of enhancing.

CBDRR has been emphasized because local communities have been weakened. The three cases in this chapter highlight some lessons learned that could enhance the adaptative capabilities of communities in East and Southeastern Asian countries.

First, the Katahira case described a model for developing sustainable CBDRR. This case positioned DRR activities as part of a community development plan that motivated children to participate in community activities by institutionalizing a children's group in the community development structure, organizing 'fun' events for children, such as the DRRTHG, and demonstrating the devotion of local adults to their community. Fostering the next generation to work for community development is key to sustainable CBDRR.

Second, the case of Ishinomaki's school-based DRR education program described one of the approaches being taken to understand local disaster risks. This program is composed of several steps, including strengthening map-reading skills, understanding the actual geography of the school district, comparing topographical and hazard maps of the school district, confirming local disaster risks by town-watching, and producing DRR maps to present the findings. These are useful steps for both children and adults to realize collaboration between schools and communities in examining local disaster risks and producing original DRR maps.

Third, the case in Joso City demonstrated the importance of facilitating daily communication within the community to ensure smooth communication in the event of a disaster. Although wide variation is seen in local communities among regions and countries, most have deteriorated along with modernization, as more people consider community ties to be annoying. However, in a world in which climate change is accelerating and catastrophic disasters are expected to occur increasingly frequently, the local community is more important than ever. Of course, in that case, the local community is not simply a thing of the past; it is a hybrid type of community that includes, for example, encouraging intergenerational communication through various community development activities, facilitating mutual learning about the local community between local residents and schoolchildren, and building new communication networks via the use of SNS. Communities in different regions and countries should therefore pursue their ideal form according to actual local conditions.

References

International Collaborating Center of Disaster Education Research and Implementation, Tohoku University (2018) The Practical Guide to a "Reconstruction-Disaster Risk Reduction Mapping-Understanding the natural environment and the lives of those in your hometown. http://drreducollabo.sakura.ne.jp/cms/wp-content/uploads/jissen_no_tebiki_en.pdf

Sakurai A (2021) School safety management: international framework and Japanese practice. In: Jing Y et al (eds) Risk Management in East Asia. Springer Nature. (in print)

Sakurai A, Sato T (2017) Enhancing community resilience through capacity development after GEJE: the case of Sendaishi-chiiki Bousai leaders (SBLs) in Miyagi prefecture. In: Santiago-Fandiño V, Sato S, Maki N, Iuchi K (eds) The 2011 Japan earthquake and tsunami: reconstruction and restoration: insights and assessment after 5 years. Springer, pp 113–126

Sakurai A et al (2019) Linking geomorphological features and disaster risk in a school district: the development of an in-service teacher training programme. IOP Conference Series: Earth and Environmental Science, Vol 630, The 12th Aceh International Workshop and Expo on Sustainable Tsunami Disaster Recovery 7–8 November 2019, Sendai, Japan

Sakurai A et al (2020) Impact evaluation of a school-based disaster education program in a city affected by the 2011 great East Japan earthquake and tsunami disaster. Int J Disaster Risk Reduct 47. https://doi.org/10.1016/j.ijdrr.2020.101632Get

Sato T, Sakurai A, et al (2018) Sustainable community development for disaster resilience and human resource development for disaster risk reduction- KAHATIRA-style disaster resilient community development. J Disaster Res 13(7):1288–1297

Sato T, Sakurai A, et al (2020) Sustainable community development for disaster resilience and human resource development for disaster risk reduction: growth and community contribution of the Katahira Children's Board for Community Development. J Disaster Res 15(7):931–942

Satomura S, Sutou J, Itou K, Hiraide R, Kandatsu T, Mizokami H, Kobayashi H, Kawashima H, Shirakawa N, Ito T, Tomioka H, Ayukawa K (2020) Social experiment for my-timeline development to improve residents' awareness of flood disaster prevention. J JSCE 8(1):261–273

Shaw R (2016) Community-based disaster risk reduction. Oxford Research Encyclopedia of Natural Hazard Science. https://doi.org/10.1093/acrefore/9780199389407.013.47

United Nations Office for Disaster Risk Reduction (2015) Sendai framework for disaster risk reduction 2015–2030. UNISDR, Geneva

Wagamachi Neshinden (My hometown Neshinden). http://neshinden.com/hotmail (in Japanese)

Chapter 9
The Practice of Education for Disaster Risk Reduction in Vietnam: Lessons Learned from a Decade of Implementation 2010–2020

Thi Thi My Tong, Duong Thi Hong Nguyen, Hung The Nguyen, and Tae Yoon Park

9.1 Introduction

The increasing damage caused by natural disasters, particularly climatic disasters such as typhoons, floods, droughts, and heatwaves, threatens the development of all economic sectors in Vietnam. In particular, the impact of natural disasters on the education sector affects thousands of students and teachers, because it interrupts their education and significantly reduces educational quality. At the same time, education is increasingly being viewed as more essential for mitigating risk and for strengthening people's capacity to respond to disasters. Disaster risk reduction education (DRRE) initiatives have become more widely adopted among international and national organizations. The importance of disaster risk reduction (DRR) is currently a major concern of the United Nations, and has resulted in the establishment of numerous initiatives such as the International Decade for Natural Disaster Reduction, the International Framework for Action for the International Decade for Natural Disaster Reduction, the Yokohama Strategy, the Plan of Action for a Safer World adopted by the 1st World Conference on Natural Disaster Reduction (UNISDR 2004), the International Strategy for Disaster Reduction, and the adoption of the Hyogo Framework for Action at the 2nd World Conference on Disaster Reduction in 2005 (UNISDR 2005). The role of education was highlighted as an

T. T. M. Tong (✉) · D. T. H. Nguyen
Vietnam Institute of Economics, Vietnam Academy of Social Science, Hanoi, Vietnam

H. T. Nguyen
Department of Climate Change and Sustainable Development, Hanoi University of Natural Resource and Environment, Hanoi, Vietnam

T. Y. Park
Graduate School of Education, Yonsei University, Seoul, South Korea

© The Author(s) 2022   101
T. Ito et al. (eds.), *Interlocal Adaptations to Climate Change in East and Southeast Asia*, SpringerBriefs in Climate Studies,
https://doi.org/10.1007/978-3-030-81207-2_9

important strategy for achieving the four priorities set out in the Sendai Framework for Disaster risk reduction 2015–2030 (UNFCCC 2015).

Vietnam was ranked sixth among countries most affected by extreme climatic events in 1999–2018 by the Global Climate Risk Index (David Eckstein 2019). The country frequently experiences severe and unpredictable climatic events, and it is estimated that these will likely become worse under the RCP4.5 scenario as the average temperature, rainfall and sea-level rise increase by 0.1 °C, 5–15%, and 20%, respectively (MONRE 2016).

With the proportion of children making up almost a quarter of Vietnam's population, an increase in the intensity of natural disasters and climate change will have a considerable impact on their development. Children are always the group most affected by natural disasters due to their physical and psychosocial vulnerability, as well as their lack of access to essential services such as health and education. In this context, DRRE is considered to be an essential strategy for sustainable development in the long-term. DRRE initiatives will better prepare children, families, and communities for shocks and significant recovery. Consequently, investing in risk reduction measures in schools is considered to be fundamental to social growth, development, and cohesion in Vietnam and for achieving the Sustainable Development Goals.

Since the beginning of the 2010s, numerous education projects on climate change response and DRR have been conducted throughout Vietnam. Most of these projects promote two main initiatives; making schools safer, and mainstreaming climate change and DRR into the school curriculum. In 2011, the Ministry of Education and Training (MOET) published the *Action Plan on implementing National Strategy for Natural Disaster Prevention, Response, and Mitigation to 2020* (MOET 2011). This plan has facilitated the development of content on climate change and DRR for use in the national curriculum of public schools. In 2014, the Ministry of Education approved the project "Information and propaganda on climate change response and natural disaster prevention and control in schools in the period 2013-2020". In 2016, specific guidance for carrying out DRRE in schools was developed and applied nationwide (MOET 2016).

Within this context, this study attempts to review the DRRE initiatives over one decade of implementation (2010–2020). The findings will clarify the lessons learned from DRRE by various stakeholders, including government and non-governmental organizations.

9.2   DRRE Initiatives in Vietnam

DRRE arose out of a long history of coping with natural disasters in Vietnam. The local communities have prepared for, responded to, and recovered from natural disasters using their wisdom and knowledge, which they passed from one generation to the next. In some ways, these indigenous practices can be seen as one of the first examples of DRRE and they are considered to be a valuable source of information to communities and schools on DRR.

Before the National Action Plan on DRR was adopted by the education sector, DRRE initiatives were mainly implemented by NGOs and other organizations outside the government. According to a mapping exercise that collected information regarding projects and programs on communication and education related to climate change and DRR by Live and Learn in Vietnam (Live and Learn Vietnam 2012), more than 50% of the programs and projects were implemented by INGOs and NGOs in Vietnam (Fig. 9.1). The topic focused mainly on climate change education (52%), DRR (28%), and a combination of both themes (20%) (Live and Learn Vietnam 2012) (Fig. 9.2).

The following section will describe 14 programs and projects conducted by MOET, NGOs, and other organizations. The findings showed that DRRE in schools covered or integrated activities on school facilities/building safety, injury prevention, swimming lessons, extracurricular activities (such as disaster prevention drills or competitions), environmental protection activities, the integration of DRR into the curriculum (such as the development of reference books for students and teaching materials for teachers, and training for teachers and students), and raising awareness in the community and among parents. Projects and programs conducted by MOET and government agencies focused on the safety of school facilities/school buildings and the development of teaching and learning materials related to DRR. In addition, DRRE activities conducted by NGOs and other organizations focused mainly on extracurricular activities and raising awareness in the community and among parents (Table 9.1).

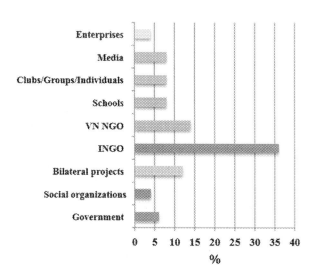

Fig. 9.1 Proportion of organizations that adopted DRRE initiatives. (Source: Live and Learn Viet Nam 2012)

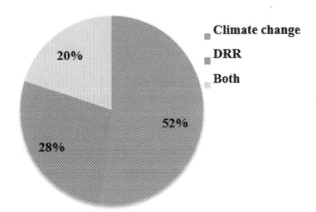

Fig. 9.2 Topics of DRRE initiatives. (Source: Live and Learn Viet Nam 2012)

9.2.1 Governmental Programs

9.2.1.1 Initiatives by the MOET

At the national level, every year, the MOET has prepared instructional materials on the preparation for, and action response to, disasters. These materials are typically issued at the beginning of the academic year or at the beginning of the disaster seasons to remind schools, educational institutions, and other educational offices about disaster prevention, response, and recovery. In 2008, the MOET initiated the Child-Friendly School program from primary to secondary level, which laid the foundation for effective DRRE implementation. One of the program's salient features was to target the provision of "fundamental learning conditions" whereby all schools are obligated to find a way to provide quality learning, especially to the most at-risk students. Accordingly, the MOET defined the following five standards for Child-Friendly Schools in Vietnam: (1) Proactively inclusive, seeking out and enabling the participation of all children, especially those who are differ ethnically, culturally, socioeconomically, and in individual ability; (2) Effective and relevant to children's needs regarding the knowledge and skills required for life and livelihood; (3) Be healthy and safe environments for, and to be protective of, children's emotional, psychological and physical well-being; (4) Gender-responsive in creating environments and capacities that foster equality; and (5) Actively engaged in, and enabling of, student, family and community participation in all aspects of school policy, management and support for children (MOET 2008).

In 2010, the MOET promoted the *Action plan for the integration of climate change into the curriculum*. The outputs of this project are (1) a set of textbooks, training and learning materials (curriculum and extracurricular activities) on climate change for all education and training levels; (2) a set of training materials for administrative officers, teachers, lecturers, students; (3) passing recommendations regarding the targets, contents, and solutions of the education sector to respond to

Table 9.1 Review of 14 DRRE projects and programs in schools carried out by various stakeholders during 2010–2020 in Vietnam

Name of projects/programs	Year	Organizations	DRR-related contents				
			(1)	(2)	(3)	(4)	(5)
Child-friendly school	2008–2009 (pilot) 2013 (scale up)	MOET	Y	N	N	N	Y
Strengthening the infrastructure capacity of schools	2008	MOET	Y	N	N	N	N
Integration of climate change into the curriculum	2010	MOET	N	Y	N	N	N
Raising community awareness and community-based disaster risk management	2009	MARD	N	N	Y	Y	Y
Communication programs using multi-media to raise awareness about climate change and DRR	2011	MOIT	N	N	N	N	Y
Introducing disaster preparedness in primary schools	2001	Vietnamese Red Cross (VNRC)	Y	Y	Y	Y	Y
Adaptation to Climate change through the Promotion of Biodiversity	2005	GIZ Vietnam	N	N	Y	N	Y
Integration of DRR into schools	2008	Development Workshop France (DWF)	N	Y	Y	N	N
Integrated disaster preparedness in Thua Thien Hue Province	2010	VNRC	N	N	Y	Y	Y
Partnership for community action on climate change	2011	The Centre for Development of Community Initiative and Environment (C&E)	N	N	Y	N	Y
Capacity building for DRR education for schools in Central Vietnam	2011–2013	SEEDs Asia	N	Y	Y	N	N
Introducing climate change and DRR education into the secondary school system in Hanoi, Quang Ninh, Hai Phong, Danang, and Ho Chi Minh City	2012	Live and Learn Vietnam	N	Y	Y	N	N
Building resilience to climate change in urban areas through integrated education	2012–2014	Institute for Environmental and Social Transition (ISET)	N	Y	Y	N	N

(continued)

Table 9.1 (continued)

Name of projects/programs	Year	Organizations	DRR-related contents				
			(1)	(2)	(3)	(4)	(5)
Capacity building for disaster risk reduction education for schools, coastal communities in Quang Nam province	2013–2015	SEEDs Asia	N	Y	Y	N	N

(1) School facilities/building safety
(2) Integration of DRR contents into the curriculum
(3) Extracurricular activities
(4) Injury prevention and swimming lessons
(4) Raising awareness- for parents/communities

Fig. 9.3 Reference books on climate change and DRR approved and published by MOET in cooperation with various organizations

climate change in the future in order to develop new education programs; and reports on the experience of the education sector on climate change in other countries. The program has successfully achieved these specific aims. The MOET has collaborated with international organizations, NGOs, research institutes, and universities, such as the United Nations Children's Fund (UNICEF), Save the Children, Oxfam, and Humanitarian Aid and Civil Protection department (ECHO), the Joint Advocacy Network Initiative(JANI), to compose teaching and learning materials for teachers and students. In 2016, the MOET issued *Guidelines for teaching climate change integration in the school curriculum*, which focused specifically on Physics, Chemistry, Biology, Geography, and Technology. As a result, 63/63 provinces and cities implemented training courses on climate change and disaster education in schools (Fig. 9.3).

In 2011, as part of the climate change adaption program, MOET had committed to carry out tasks related to DRRE, including the introduction of education programs for children, which focus on responding and adapting to impacts of natural disasters and climate change through a new education development strategy for the period 2011–2015. In 2014, the MOET approved the project "*Information and propaganda on climate change response and natural disaster prevention and control in schools in the period 2013–2020*". The target audiences of the project include children, students, staff, teachers, members of the education sector, parents, and the community. In addition, several other programs and projects related to DRRE were carried by the MOET, such as inclusion of material on environmental protection into national education systems, the program "Strengthening the infrastructure capacity of schools," educational project targeting primary school children from economically disadvantaged communities, and a project to create access to secondary education for students in remote areas.

9.2.1.2   Initiatives by Other Governmental Organizations

In 2009, the Disaster Management Center, which falls under the auspices of the Ministry of Agriculture and Rural Development (MARD), promoted the program "Raising community awareness and community-based disaster risk management" following a promulgation of resolution No. 1002/QD-TTg by the Prime Minister in 2009. The program's main objectives are to raise community awareness and effectively organize a model for community-based disaster risk management for all levels and line agencies, particularly for the local authorities and residents at the village and commune levels. In 2011, the Ministry of Information and Technology (MOIT) developed multi-media communication programs (i.e., printed newspapers, radio, TV, online newspapers) to raise awareness about climate change and DRR.

9.2.1.3   Initiatives by International Organizations and Non-governmental Organizations

In 2001, the Vietnam Red Cross Society (VNRC) emphasized disaster preparedness activities and implemented the program "Introducing Disaster Preparedness in Primary Schools." Since then, the program's activities have been replicated and are underway in all 21 of the most disaster-prone provinces in Vietnam. The aim of these programs was to mitigate the risk of disasters among school-going children who are among the most vulnerable to disasters. Targeted beneficiaries were teachers and children, as well as VNRC staff and government personnel. The program developed a new package of disaster preparedness training materials for Red Cross personnel, community leaders, teachers, and children. In addition, the program also elicited the active participation of relevant stakeholders, including teachers and children, in writing and finalizing the training and learning material (UNESCO 2009). The 12-month program aimed to assess needs and develop training materials for different targets including trainers, teachers, students (grade 4 and grade 5) and other relevant stakeholders in the selected communes. Recipient schools in disaster-prone areas also organized inter-provincial competitions, including drama, quizzes, and painting competitions based on a guidance from disaster preparedness booklet. However, the program faced significant challenges in integrating a disaster preparedness component into the official training curricula due to the risk of overburdening school children. This challenge has not yet been overcome, although the VNRC insisted on providing disaster preparedness training until 2010 to teachers and children in eight coastal provinces in northern Vietnam with financial support from the Japanese Red Cross. As a result, the program and its subsequent replication have helped train 15,000 teachers and over 500,000 school children over the last 6 years (UNISDR 2007).

In 2005, the Deutsche Gesellschaft für Internationale Zusammenarbeit (GIZ) GmbH Vietnam carried out a project titled, "Adaptation to climate change through the promotion of biodiversity," which was financed by the German Federal Ministry for Environment, Nature Conservation, Nuclear Safety. It targeted communities, school teachers, and students, and tried to raise awareness of environmental and natural resource protection and biodiversity conservation. The primary approach were infused through the curriculum, specifically in Biology, Geography, Civic Education subjects and extracurricular activities, such as the Green and Clean Day, environmental drawing competitions, bird sanctuary visits, etc. As a result, the projects benefited teachers and students from grade 1–12 at 154 primary schools, 74 secondary schools, and 19 high schools in Bac Lieu Province.

In addition, an excellent example of integrating DRR into schools was implemented by Development Workshop France (DWF) in Hue Province in 2008. DWF worked as an important partner of many communes in Hue Province to strengthen existing public infrastructure and build safe, new schools, markets, and health facilities. At the village level, DWF accompanied with communes to build safe kindergartens and to strengthen primary schools so that children could learn about safety and safe construction techniques in a safe environment and take these principles back to their families. DWF also trained teachers about disaster prevention and ran workshops with children about disaster prevention and the child's role and needs before, during, and after disasters. Children were active in school and the community in promoting the vulnerability reduction message (DWF 2009).

Other projects implemented in Hue Province included, "Integrated Disaster Preparedness in Thua Thien Hue Province" which was undertaken by the German Red Cross (GRC) in cooperation with the Vietnamese Red Cross (VNRC) in 2010, "Partnership for community action on climate change" by the Centre for Development of Community Initiative and Environment (C&E) and Global Action Plan International (GAP) in 2011. The project used schools as hubs for sustainability actions by households and communities, engaging teachers in civil society, linking schools and students with existing Vietnam government programs and priorities as well as existing Swedish government programs and priorities, which also helped to build partnerships between organizations. The activities included raising awareness among junior secondary schools' students to perform actions related to climate change, carrying out community outreach (family, school, neighbors) through undertaking Eco-team sustainable projects and activities, and establishing a "Green living model" at schools. The outcomes of the project included materials for community climate action through schools which were adopted for Vietnamese rural and urban contexts; teachers and youth leaders/facilitators received training to engage students in climate action; schools facilitated community climate activities; eco-team models were shared and advocated in project sites; and links to international development and learning about climate actions and education for sustainable development (ESD) were strengthened. Further, a household-level climate mitigation model has been developed, including teaching materials, materials for students, and a process to mainstream content into the curriculum. The project has benefited 3000 students, 500 junior secondary students, 93 teachers, 50 VNGO staff have participated and received training, and 20 education and sustainability decision-makers have been trained in Hanoi, Thua Thien Hue, and Thai Binh.

In 2012, Live and Learn cooperated with the British Council in Vietnam and made considerable efforts to introduce climate change and DRR education into the secondary school system in Hanoi, Quang Ninh, Hai Phong, Danang, and Ho Chi Minh City. In addition to developing interactive multi-media materials (video and website) on "child/youth and climate change," the project also focused on the development of educational materials which could be used for adopting climate and disaster prevention actions in schools and communities. It also included various activities, such as training resource teachers and active youth clubs on climate change education; organizing extracurricular activities for students using the educational materials with the help from the resource teachers and volunteers of youth clubs; establishing a child-led forum at schools and in communities on climate change adaptation and mitigation; running competitions for every school that took part in the project to motivate them throughout the year and to identify good initiatives to support and potentially expand. Small grants were awarded for children-led initiatives on climate change adaptation and mitigation, and British Council schools were used as an online platform for promoting different initiatives.

In Da Nang, numerous projects and programs related to DRRE were carried out in the city. The project "Building resilience to climate change in urban areas through integrated education" was funded by the Rockefeller Foundation and implemented through a collaboration between the Institute for Environmental and Social Transition (ISET) and DoET Da Nang from 2012 through 2014. This project is widely considered to be a successful case study. Briefly, it started in January 2012 at three core schools (Ngo Quyen Primary School, Nguyen Van Linh Secondary School, and Hoa Vang High School) with the aim of developing a separate teaching syllabus on climate change and DRR based on the actual local situation. The project established a group of 30 core teachers to compile a syllabus for three subjects at the primary level (Geography, Biology, and Civics education). In addition, the project also targeted extracurricular activities in many forms, such as competitions, drills.

Another project that focused specifically on DRRE at the school level was carried out in Da Nang City in 2012. The project, which was implemented by SEEDS Asia and funded by JICA, aimed to build a disaster education network among schools and related organizations to enhance school DRRE in Da Nang city. The primary outcomes of the project were: (1) Core Schools for DRR (DRR Core schools) were established in every district of Da Nang city; (2) DRR Core Schools built a network on DRR; (3) More than one teacher at each school would be capable of conducting DRR classes; (4) Teachers and related governmental officials enhanced their knowledge of DRR education in Thua Thien Hue Province and Quang Nam Province; (5) A module for active teachers was developed; and (6) Experiences and outcomes were shared among a variety of disaster-related organization staff. As a result, seven core DRR schools were selected, 88 teachers were trained to become Training of trainers (ToT), 357 lessons and learning DRR classes were organized, and a working group comprising members from a variety of organizations (including MOET, BOET, vice-principal, and teachers of DRR core schools) was established (Ueda et al. 2016).

Other organizations from social organizations, private organizations, academia, and media have also contributed to the promotion of DRRE. In 2010, the Youth Union of the Information Technology Department at Hanoi University developed the software program "Awareness education for youngsters on climate change responses," which can be accessed both online and offline. This is an excellent resource for disseminating knowledge about climate change and disaster impacts for Youth Union members. The website "What can youngsters do about climate change?" provides activities, such as writing and drawing competitions with content reflecting the methods and responsibilities of individuals and organizations in response to climate change and disasters. Interestingly, the integration of climate and disaster issues has also been incorporated into the Mathematics curriculum. The project had been carried out by Kien Hung Secondary in Ha Noi with financial support from the World Bank. The project provided about 40 mathematics questions that were related to climate change topics. The results from the project have been shared with other schools and replicated in other provinces.

9.3 Lessons Learned from a Decade of DRRE Implementation in Vietnam

This review and analysis of the activities of 14 projects implemented in the period 2010–2020 in Vietnam have shown that many lessons related to the cohesion and role of local government have been learned through sustainable and innovative design initiatives.

One of the first lessons is that close cooperation between the local authorities and the Department of Education and Training during project implementation has been critical to a project's success. The project "Building resilience to climate change in urban areas through integrated education" that was implemented by ISET established a close relationship between the project management board, the local government, and the Da Nang Department of Education. This close relationship played a central role in facilitating and encouraging schools to participate both actively and proactively to achieve the project's aims. Furthermore, thanks to the involvement of the local government and the Department of Education and Training from the project design stage, the project received considerable support from government organizations throughout its implementation and the results of the project have been widely shared. Importantly, the lessons of the project have been reviewed thoroughly and many aspects have been integrated into the policies of local authorities to promote local DRR activities.

Secondly, it is clear that those projects that consider sustainability from the outset, i.e., from the project design stage, will achieve greater efficiency and sustainability. With the establishment of DRR core schools and working groups, the projects have promoted a strong connection between all stakeholders, creating space for them to learn from each other and to share what they have learned; indeed, such a connection is the key to the sustainability of a project. For example, although SEEDS Asia's project ended more than 5 years ago, some of the project's activities are still being implemented, especially at DRR core schools.

Another lesson is about incorporating indigenous knowledge when implementing the DRRE project. Vietnam has a long history of responding to disasters, and the local communities have prepared, responded, and recovered from these natural disasters using their experience and knowledge, which they have passed from one generation to the next. Indeed, these indigenous practices could be considered to be one of the first DRRE initiatives, and they are a valuable source of information and make an important contribution to both the community and school-based education initiative related to DRR.

Acknowledgement This research is funded by Vietnam National Foundation for Science and Technology Development (NAFOSTED) under grant number 507.01-2019.302.

References

David Eckstein VK (2019) The global climate risk index 2020, who suffers Most from extreme weather events? Weather-related loss events in 2018 and 1999 to 2018. Germanwatch, Berlin

DWF (2009) Developing Access to Safe and Good Quality Schools and Health Centres in Rural Central Viet Nam

Live and Learn Viet Nam (2012) Mapping climate change and disaster risk reduction education and communication in Viet Nam. Live and Learn, Plan Viet Nam, AID Australia

MOET (2008) Directive No. 40/2008/CT-BGDĐT dated 22/7/2008 on launching the emulation movement to build friendly schools, active students in high schools in the period 2008–2013. Ministry of Education and Training, Hanoi

MOET (2011) The action plan to implement the National Strategy on natural disaster mitigation, prevention and control in the education sector. Ministry of Education and training, Hanoi

MOET (2016) Guidelines for teaching climate change integration in the school curriculum. Ministry of Education and Training, Hanoi

MONRE (2016) Climate Change and Sea Level Rise Scenarios for Vietnam. https://www.preventionweb.net/files/11348_ClimateChangeSeaLevelScenariosforVi.pdf

Ueda Y, Matsumoto E, Nakagawa Y, Shaw R (2016) International cooperation: grassroots experience sharing in Vietnam. https://doi.org/10.1007/978-4-431-55982-5_13

UNESCO (2009) Policy guidelines on inclusion in education. Paris, France. https://unesdoc. unesco.org/ark:/48223/pf0000177849

UNFCCC (2015) Adoption of the Paris Agreement, vol 21932. United Nations Framework Convention on Climate Change. http://unfccc.int/resource/docs/2015/cop21/eng/l09r01.pdf

UNISDR (2004) Living with risk. A global review of disaster reduction initiatives, Strategy, vol 1. United Nations Inter-Agency Secretariat of the International Strategy for Disaster Reduction, Geneva

UNISDR (2005) Knowledge, innovation and education: building culture of safety and resilience. In: World conference on disaster reduction. United Nations/International Strategy for Disaster Reduction, Kobe

UNISDR (2007) Hyogo framework for action 2005–2015: Building the Resilience of Nations and Communities to Disasters. http://www.unisdr.org/files/1037_hyogoframeworkforac- tionenglish.pdf

Part III
Conclusion

Chapter 10
Sharing Interlocal Adaptation Lessons

Makoto Tamura and Tetsuji Ito

10.1 Sharing Lessons from Case Studies

As seen in the previous chapters, Asia is one of the regions most vulnerable to the impacts of climate change, as it represents more than 60% of the world's population, making it the growth center of the world. The climate in East and Southeast Asia is incredibly diverse and includes many climate zones, from tropical to subarctic. The impacts of and adaptations to climate change is this region are also diverse. A variety of adaptation practices have already employed to minimize the risk of climate change among the countries in this area, and a number of similarities and ongoing challenges are apparent. Therefore, to adapt more effectively to climate change, it is essential for these countries to share their knowledge and experiences with each other.

Table 10.1 summarizes the impacts and adaptations discussed in this book, which mainly deals with "agriculture and natural resource management" and "disaster risk reduction and human resource development". This book discusses numerous impacts and diverse adaptations measures, both hard and soft. The impacts on these two sectors are classified in Table 10.1, with some overlap.

Table 10.2 shows some of the extracted lessons discussed in this book. The following commonalities and implications were observed:

M. Tamura (✉)
Global and Local Environment Co-creation Institute, Ibaraki University, Ibaraki, Japan

VNU Vietnam Japan University, Hanoi, Vietnam
e-mail: makoto.tamura.rks@vc.ibaraki.ac.jp

T. Ito
College of Humanities and Social Sciences/Global and Local Environment
Co-creation Institute, Ibaraki University, Ibaraki, Japan

© The Author(s) 2022
T. Ito et al. (eds.), *Interlocal Adaptations to Climate Change in East and Southeast Asia*, SpringerBriefs in Climate Studies,
https://doi.org/10.1007/978-3-030-81207-2_10

Table 10.1 Summary of impacts and adaptations discussed in this book

	Impacts	Adaptations
Agriculture and natural resource management	Water scarcity (Pulhin et al.; Vongtanaboon)	Participatory water (or watershed) management
	Crop production (Ishikawa & Furuya; Matsuura & Sakagami)	Projection, changes in planting dates and cultivars
	Food supply and markets (Ishikawa & Furuya)	System of rice intensification (SRI)
	Ecosystem service depletion (Pulhin et al.; Balderama; Bengen et al.; Oktarina et al.)	Different tillage practices or crop rotations
	Mangrove forest decrease and fishery resource (T.H.Nguyen)	High temperature tolerant rice cultivars <= > market impacts (e.g., price)
		Information communication technology (ICT). Smart agriculture system for corn and dialogical tools such as GIS and 3D mapping
		Stakeholder involvement, e.g., fishery livelihood, oil palm biomass utilization for climate smart agriculture
		(Young) human resource development
Disaster risk reduction and human resource development	Sea level rise (SLR) and erosion (Yasuhara & Murakami; Tamura & Pham; Q.V. Nguyen)	Monitoring the current situation (e.g., SLR, coastal change, LS)
	Land subsidence (LS; Yasuhara & Murakami)	Projection = > "reactive" to "proactive" adaptation
	Coral reefs (Bengen)	Zoning based conservation (e.g., mangrove planting and Casuarina trees outside of mangrove)
	Mangrove decrease (Bengen; Q.V. Nguyen)	Hard and soft protection (e.g., dike, mangrove, and multiple protection, and early warning)
	Flooding and disasters (Pulhin et al.; Balderama; Sakurai & Ito; Tong et al.)	Watershed management (participatory approach and ICT)
	Earthquake and Tsunami (Sakurai & Ito)	Sustainable livelihood
		Community-based disaster risk reduction (CBDRR)
		Disaster risk reduction education (DRRE)
		Raising awareness => Stakeholder involvement

Table 10.2 Some lessons learned in the book.

Part I: Agriculture and natural resource management	
Participatory climate change adaptation using watershed approach: Processes and lessons from the Philippines	1. Recognizing the different scales of adaptation through a watershed approach
	2. A participatory framework for catalyzing collective action among stakeholders
	3. Solution-based analysis that incorporates local knowledge
	4. Community-based adaptation needs to be effectively linked to higher scales of governance to enhance the resilience of communities and ecosystems
Climate change adaptation practices towards sustainable watershed management: The case of Abuan watershed in Ilagan City, Philippines	A good example of watershed management is how to align and plan the interface of a project framework to involve local governments and facilitate its effective adaption among stakeholders.
Economic evaluation of adaption measure of rice production in Vietnam with supply and demand model	A supply and demand model for rice production was developed for an economic evaluation of adaptation. Climate change adaptation measures for continued rice production, such as adopting short-duration or salinity-tolerant cultivars, have been introduced and evaluated.
Coastal-small island ecosystems and conservation perspectives within adaptation efforts (*in Indonesia*)	Sustainable small coastal island conservation management requires thorough participatory planning. The design requirements may include:
	1. A complete set of data and information
	2. Evaluation of program relevance in relation to the conditions in particular areas
	3. Evaluation of program impacts on the environment and communities
	4. Involvement of stakeholders and communities in planning
	5. Dissemination of information strategies
Part II: Disaster risk reduction and human resource development	
Geotechnical approaches to disaster risk reduction in Japan and Vietnam	1. Inundation of coastal regions undergoing a combination of sea level rise and land subsidence (LS) is more serious than in regions without LS. Multiple adaptive measures can be adopted.
	2. The concept of adaptation should be shifted from being reactive to proactive, as doing so would lead to increased overall resilience.

(continued)

Table 10.2 (continued)

Developing sustainable and climate-resilient livelihood in Giao An commune, Giao Thuy district, Nam Dinh province (*in Vietnam*)	For human capital:
	1. Organizing short- and long-term training courses depending on livelihood characteristics to equip local residents with the necessary knowledge and skills to carry out production activities.
	2. Organizing community events combined with contests related to climate change
	For social capital:
	1. Studying how to create more effective production groups, teams, and areas
Community-based disaster risk reduction education in Japan	1. CBDRR is at the core of any type of risk reduction.
	2. Local disaster risk must be understood for the determination of measures to reduce disaster risk in the community.
	3. In Japan, expectations of enhancing CBDRR have been emphasized while local communities have been weakened. Three cases of disaster risk reduction (DRR) in Japanese schools and communities provide some lessons to enhance adaptive capacities.
The practice of education for disaster risk reduction in Vietnam: Lessons learned from a decade of implementation 2010–2020	Review of DRRE in Vietnam
	1. Close cooperation between local authorities and the Department of Education during project implementation has been critical.
	2. Projects that consider sustainability from the outset, i.e., from the project design stage, will achieve greater efficiency and sustainability.
	3. Incorporating indigenous knowledge into DRRE projects is valuable.

1. Climate change impacts are already evident in many regions and sectors, as shown in Table 10.1. People are adapting to current climate risks and need to adapt to future risks at the same time.

2. Two main approaches to climate adaptation have been developed to address adaptation: a top-down scientific approach and a bottom-up regional approach, as discussed in Chap. 1. The scientific approach can contribute to adaptation strategies such as monitoring, projection, impact/vulnerability assessments, and adaptation policy planning. These cycles are periodically checked and implemented.

3. The regional approach, which involves collaboration among experts and stakeholders and incorporates traditional and indigenous knowledge, can enhance adaptive capacities for communities and regions. Because the impacts of climate change, vulnerabilities, and priorities vary substantially depending on regional characteristics, both regional and scientific approaches can support information sharing in relation to decision making.

Stakeholder involvement and collaboration with national and local institutions are required for adaptation. This book covers both adaptation research and activities among some East and Southeast Asia countries to fill the gaps between scientific and regional approaches. Some tools or methods for enhancing this collaboration include a participatory framework, community-based adaptation, and education in relation to resource management and disaster risk reduction. In this context, local and indigenous knowledge plays an important role in the formulation of adaptation governance and related strategies (IPCC 2007).

10.2   Discussion of the Three Aims of This Book

This book had the following three aims, as mentioned in the preface.

1. Promote interlocal lessons learned by sharing climate change adaptations, such as through "agriculture and natural resource management" and "disaster risk reduction and human resource development"
2. Develop new adaptation measures and research approaches that can consider the regional nature of East and Southeast Asia
3. Share practical adaptation options that have permeated society in each country/region

Regarding 1), the lessons learned and rooted in each country or region are drawn as summarized above. However, promoting them "interlocally" is still required as a next step. For example, participatory climate change adaptation using the watershed approach reported by Philippine researchers could be practically applied in other countries and regions. However, it is necessary to consider not only the differences in the natural environment and ecosystem of the area, but also the customs and culture of the specific procedure, including how residents and stakeholders can participate. If the place changes, it is natural that the method must also be changed.

Needless to say, the researchers, local governments, and citizens in each country or region must be careful about this point when attempting to apply the practices of other countries or regions to their own. This book does not always provide concrete examples of how to apply the climate change adaptation practices reported here in each country or region to other places. We think that it needs to be frankly acknowledged that this book is only intended to be a catalyst for doing so.

In the preface, we mentioned that the term "interlocal", which we describe as the concept of connecting localities, transforming points into lines, lines into planes, and finally, planes into three-dimensional objects (Fig. 10.1), is not very familiar in English (Yamori and Ito 2009). However, the range of locality needs to be clarified. For example, even in Japan, the climate and environment in Hokkaido in the north differ from those in Okinawa in the south, as do ways of thinking among residents. Ibaraki University, where the editors work, is located in Ibaraki Prefecture, north of Tokyo, where agriculture is thriving; however, this locality also differs from Hokkaido and Okinawa. One locality on a country-by-country basis may not be

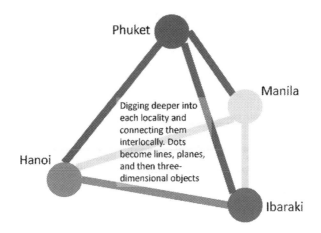

Fig. 10.1 Conceptual diagram of "interlocal"

appropriate, even in a relatively small country like Japan. Therefore, further discussion is needed on this point.

Regarding 2), it can be said that the purpose of developing new adaptation measures and research approaches has been achieved to a certain extent, in that advanced cases have been reported in each country and region. The development of an economic evaluation of adaptation measures related to rice production in Vietnam corresponds to a specific example, and other examples can be given. However, this book covers only a small part of the vast number of specific themes that should be covered on this subject.

This book deals with two major themes, "agriculture and natural resource management" and "disaster risk reduction and human resource development". Of course, we intend to deal with a wide range of concrete themes as much as possible. However, although it seems that climate change adaptation requires lifestyle changes, we have not been able to deal with how to bring about such changes. We think that the COVID-19 pandemic that continues as of 2021 will be one of the major triggers for a review of our society, life, and economy. This is one of the challenges for the near future.

Of course, 3) "sharing practical adaptation options" will be left to practice in the future. Here, we wanted to use the term "social permeation" rather than the more often used "social implementation". This does not mean that we deny "social implementation". However, it reminds us of only the top-down approach. If we desire that climate change adaptation measures be truly accepted and established, we should also value indigenous knowledge and discover what we can learn using a bottom-up approach. One option to emphasize such a point is to think about "social permeation", that is, what is needed for the policy to permeate among people.

Whether it is "social implementation" or "social permeation", we often encounter social dilemma situations when attempting to change society. Even if we know that we can benefit from working together in the long term, it can be detrimental in

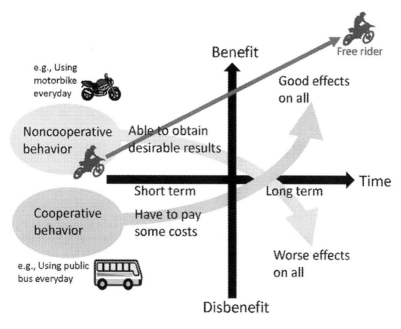

Fig. 10.2 Basic structure of the social dilemma

the short term. The emergence of "free riders", who act uncooperatively while benefiting from almost everyone else working cooperatively, complicates issues even further (Fig. 10.2). The dilemma of "tragedy of the commons" is also widely known. The issue is how to develop human resources who think about problems from a broader perspective.

10.3 A Short Note on "Adaptation"

The term "adaptation" was originally used in biology (Ito 2018). In the process of biological evolution, the term "adaptation" is used in the context of "natural selection" in relation to adapting to changes in the environment. After that, "adaptation" came to be used in psychology and sociology, but even in those fields, it basically means that living things (human beings) change according to changes in the environment. For example, when we mention "adaptation to school", we are referring to psychological changes among students (changing humans), not changes in the school (changing the environment).

In the context of climate change, adaptation is defined as "the process of adjustment to actual or expected climate and its effects. In human systems, adaptation seeks to moderate or avoid harm or exploit beneficial opportunities. In some natural systems, human intervention may facilitate adjustment to expected climate and its effects" (IPCC 2014, p.1758). The object of climate adaptation includes both changing society and human systems.

10.4   The Way Forward

This book has introduced ideas regarding both research and activities for climate change adaptation among some East and Southeast Asia countries, such as Vietnam, Thailand, Indonesia, Philippines, China, and Japan. As the impacts of climate change are expected to be increasingly serious, many promising activities and projects for climate adaptation have been proposed. These adaptations can be shared interlocally among countries, regions, and residents.

To enhance the adaptive capacities of climate change, it is critical to establish a social system that can promote information collection and the sharing and raising of awareness regarding the importance of these activities. In addition, this dialogue must be promoted among researchers in diverse fields. Adaptation to climate change should contribute to other social goals, such as the mitigation of climate change and the creation of an environmentally safe, friendly, and secure society in accordance with the Sustainable Development Goals established by the United Nations General Assembly.

It cannot be said that aims of this book have been achieved through its publication alone, and some problems still remain. However, there is no doubt that this book can contribute to the creation of new opportunities to tackle these challenges.

References

IPCC (2007) Climate change 2007 -impacts, adaptation and vulnerability: working group II contribution to the fourth assessment report of the IPCC. Cambridge University Press, London
IPCC (2014) Climate change 2014 -impacts, adaptation and vulnerability: working group II contribution to the fifth assessment report of the IPCC. Cambridge University Press, London
Ito T (2018) Prospects for a research dialogue toward the world: considering the cross-border concept of "adaptation". Jpn J Dev Psychol 29:189–198. (in Japanese)
Yamori K, Ito T (2009) Consideration of "inter-locality" using an epistolary style. Jpn J Qualit Psychol 8:43–63. (in Japanese)

Column 1
The Southeast Asia Research-Based Network on Climate Change Adaptation Science (SARNCCAS) -Weaving the Wisdom of Climate Change Adaptation into Glocal Networks-

Tetsuji Ito and Akihiko Kotera

In the twenty-first century, the problems caused by climate change are becoming increasingly apparent, especially in developing countries such as those in Southeast Asia. As scientists, we have been hoping to strengthen our academic relationship on this thesis to cooperate and fight against such global problems. Nevertheless, the effects of climate change are not the same in any region of the world. Therefore, attention must be paid to each individual, and it is vital to connect them interlocally. If we could relativize the problems of each country and region, and analyze and examine them more objectively from different angles by networking such research bases, we would likely be able to produce significant results in regard to the sustainable development of this region.

We spent a lot of time fostering these ideas among our colleagues from Vietnam, the Philippines, Thailand, and Indonesia until finally launching our network project in late 2018. This ambitious network, named "The Southeast Asia Research-based Network on Climate Change Adaptation Science (SARNCCAS)," aims to network researchers from various disciplines in Southeast Asian countries and Japan to encourage different voices toward a common message for climate change adaptation science and contribute to the research base and community-based research (Fig. C1.1). The next generation of young researchers and students also plays an essential role in this network.

Like other countries, Japan also has been strongly affected by climate change in recent years, for example, by experiencing heavy rainfall disasters. Hence Japan has

T. Ito (✉)
College of Humanities and Social Sciences/Global and Local Environment Co-creation Institute, Ibaraki University, Ibaraki, Japan
e-mail: testsuji.ito.64@vc.ibaraki.ac.jp

A. Kotera
Global and Local Environment Co-creation Institute, Ibaraki University, Ibaraki, Japan

© The Author(s) 2022
T. Ito et al. (eds.), *Interlocal Adaptations to Climate Change in East and Southeast Asia*, SpringerBriefs in Climate Studies,
https://doi.org/10.1007/978-3-030-81207-2_11

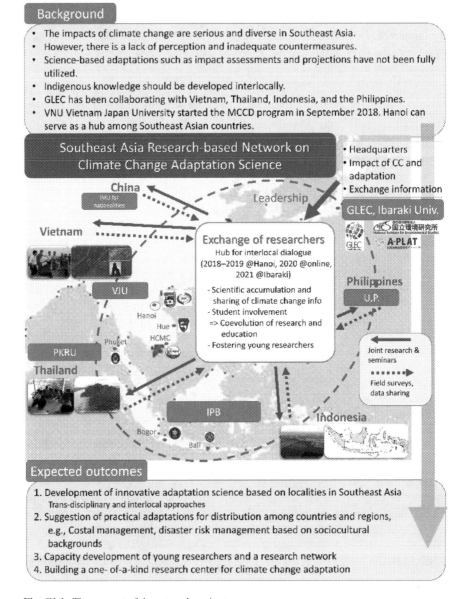

Background
- The impacts of climate change are serious and diverse in Southeast Asia.
- However, there is a lack of perception and inadequate countermeasures.
- Science-based adaptations such as impact assessments and projections have not been fully utilized.
- Indigenous knowledge should be developed interlocally.
- GLEC has been collaborating with Vietnam, Thailand, Indonesia, and the Philippines.
- VNU Vietnam Japan University started the MCCD program in September 2018. Hanoi can serve as a hub among Southeast Asian countries.

Southeast Asia Research-based Network on Climate Change Adaptation Science
- Headquarters
- Impact of CC and adaptation
- Exchange information

GLEC, Ibaraki Univ.

China
Vietnam
Leadership

Exchange of researchers
Hub for interlocal dialogue
(2018–2019 @Hanoi, 2020 @online, 2021 @Ibaraki)
- Scientific accumulation and sharing of climate change info
- Student involvement
=> Coevolution of research and education
- Fostering young researchers

VJU
Hanoi
Hue
HCMC
Phuket
PKRU
Thailand
IPB
Bogor
Bali
Indonesia

Philippines
U.P.

Joint research & seminars
Field surveys, data sharing

Expected outcomes
1. Development of innovative adaptation science based on localities in Southeast Asia
   Trans-disciplinary and interlocal approaches
2. Suggestion of practical adaptations for distribution among countries and regions, e.g., Costal management, disaster risk management based on sociocultural backgrounds
3. Capacity development of young researchers and a research network
4. Building a one- of-a-kind research center for climate change adaptation

Fig. C1.1 The concept of the network project

been working on countermeasures as an advanced problem-solving country from the early stage. Sharing these experiences can be expected to play a beneficial role in networking activities. This concept can be applied to other network member countries as well. Unlike research institutes in Western countries that are trying to do the same thing, Southeast Asian countries and Japan could share their lessons learned under a common foundation of the East. SARNCCAS, aiming to construct a network between all Southeast Asian countries, will also focus on the commonality and diversity of cultures and ways of thinking in Asia.

In the time it has taken for this book to be published, we held international workshops in Hanoi, Vietnam, in 2018 and 2019 (Fig. C1.2). Furthermore, we found that the network can continued to function robustly even under the challenging circumstances of the Coronavirus 2019 pandemic. Network members were able to work together online that incorporated new approaches such as virtual fieldwork for everyone online in 2020 (Fig. C1.3). Each seminar was attended by about 100 researchers and students, who shared their own lessons learned with everyone.

"Sharing is Learning" is the core message of SARNCCAS, which means listening to, appreciating, respecting, and learning from each individual and the entire group. This book is one of the achievements of this networking. We expect to continue to generate more interlocal knowledge in the future while making use of these new connections.

Fig. C1.2 The first workshop held at the VNU Vietnam Japan University, Vietnam, in December 2018

Fig. C1.3 Virtual fieldwork tour given at the third workshop, which was held online, in November 2020

Column 2
Variations of Short-Lived Climate Pollutants in Hanoi, Vietnam

Tung Duy Do and Kazuyuki Kita

Emissions of air pollutants have been increasing significantly in Asian countries due to the rapid development of industry and economy. Long-range, transboundary transport of these pollutants probably affects the atmospheric environment and the regional climate in this region (Kita et al. 2009). Climate change, air pollution, and sustainable development are inter-linked, and co-benefits of cutting short-lived climate pollutants (SLCP) will avoid global warming higher than 1.5 °C and negative trade-offs (CCAC 2019; IPCC 2018). Therefore, identification of SLCP emission/ production/ transportation sources is critical for planning mitigative measures to reduce SLCP.

In this study, simultaneous observation of black carbon (BC), tropospheric ozone (TO_3) and particulate matter 2.5 ($PM_{2.5}$), which are significant climate forcers, was carried out at VNU Vietnam Japan University in Hanoi to clarify the concentrations and variations of SLCP in Hanoi and Northern Vietnam. The research applied HYSPLIT trajectory model to distinguish contribution source regions of SLCPs to Hanoi and deployed remote $PM_{2.5}$ stations surrounding Hanoi and coastal region in Northeast sector of Northern Vietnam to compare upwind and downwind concentrations.

Figure C2.1 shows the maximum episodes of BC and $PM_{2.5}$ observed in wintertime, especially in January with periods lasting from 1 day to 1 week. Monthly averaged concentrations of BC, TO_3 and $PM_{2.5}$ were in range of 1–3 µg/m³, 20–50 ppbv and 18–65 µg/m³, respectively (Fig. C2.2). BC concentration was estimated

T. D. Do
VNU Vietnam Japan University, Hanoi, Vietnam

K. Kita (✉)
Graduate School of Science and Engineering, Ibaraki University, Ibaraki, Japan
e-mail: kazuyuki.kita.iu@vc.ibaraki.ac.jp

© The Author(s) 2022
T. Ito et al. (eds.), *Interlocal Adaptations to Climate Change in East and Southeast Asia*, SpringerBriefs in Climate Studies,
https://doi.org/10.1007/978-3-030-81207-2_12

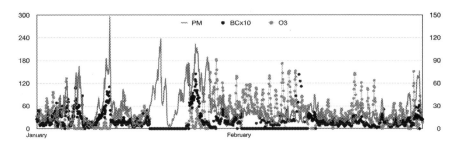

Fig. C2.1 Timeseries of BC, TO₃ and PM₂.₅ in Hanoi in Jan. and Feb., 2019

from 4% to 6% of PM2.5 in all seasons of 2019. Diurnal variations of these species suggested that major part of them were emitted or produced in Northern Vietnam region. BC and $PM_{2.5}$ were remarkably increased during rush hours or night-time in diurnal variation. In contrast, TO_3 was often high at noon and depleted to zero at night. Seasonal variation as shown in Fig. C2.2 indicated that BC and $PM_{2.5}$ increased with winter monsoon, and TO_3 actively produced in summer, indicating that air transport in association with the winter monsoon affected concentration of BC and $PM_{2.5}$ in this region. Due to observed enhances of BC and $PM_{2.5}$ in 2019, the comparison analysis with local and regional transport features focused on winter-time (Fig. C2.3). These high rises were mostly associated with trajectories from South China Sea, and detailed analyses of relation between these rises and the calculated trajectory routes revealed that these rises were actually attributed to emissions from North East coastal region of Northern Vietnam.

In REAS inventory data set (Kurokawa and Ohara 2020), national BC emissions in Vietnam was estimated as 59 Tg/y in 2015, second largest in ASEAN countries. Given the significant climate forcing of BC, this study strongly suggests that mitigation measures to reduce BC in Vietnam can considerably improve both regional climate change and air quality in the Northern Vietnam region.

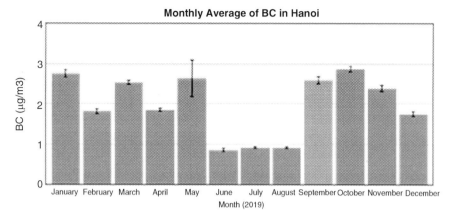

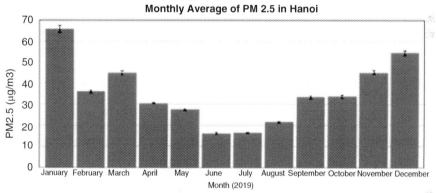

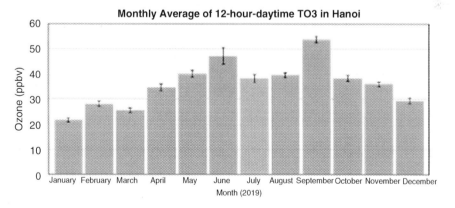

Fig. C2.2   Monthly average of BC, $PM_{2.5}$ and TO_3 in Hanoi in 2019

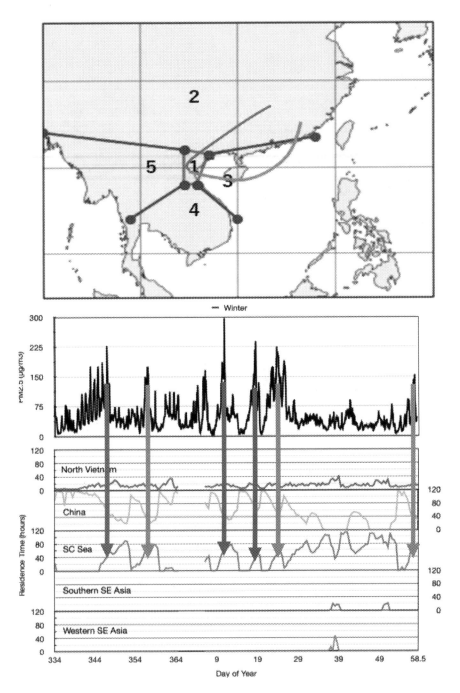

Fig. C2.3 Relation of PM$_{2.5}$ concentration with air mass transport trajectories and residence time in areas around Northern Vietnam

References

CCAC (2019) Climate and clean air coalition to reduce short lived climate pollutants. (2019, March 15). http://www.ccacoalition.org/en/content/short-lived-climate-pollutants-slcps

IPCC (2018) Global warming of 1.5 °C. An IPCC Special Report on the impacts of global warming of 1.5 °C above pre-industrial levels and related global greenhouse gas emission pathways, in the context of strengthening the global response to the threat of climate change, sustainable development, and efforts to eradicate poverty (Masson-Delmotte V, Zhai P, Pörtner HO, Roberts D, Skea J, Shukla PR, Pirani A, Moufouma-Okia W, Péan C, Pidcock R, Connors S, Matthews JBR, Chen Y, Zhou X, Gomis MI, Lonnoy E, Maycock T, Tignor M, Waterfield T (eds)). In Press

Kita K, Kasai Y, Sagi K, Hayashida S, Irie H, Kanaya Y, Miyazaki K, Takigawa M, Noguchi K, Kondo Y, Koike M, Akimoto H (2009) Transboundary air pollution in East/Southeast Asia and geostationary measurement. AGU Fall Meeting Abstracts

Kurokawa J, Ohara T (2020) Long-term historical trends in air pollutant emissions in Asia: Regional Emission inventory in ASia (REAS) version 3. Atmos Chem Phys 20:12761–12793. https://doi.org/10.5194/acp-20-12761-2020

Column 3
Mixing Grey and Green Infrastructures for Coastal Adaptation in Vietnam

Makoto Tamura and Oanh Thi Pham

Vietnam has approximately 5,600 km of coastline and is among the countries most vulnerable to future sea level rise (SLR). In terms of risk of inundation and affected population, Vietnam is one of the countries most at risk from climate change (MONRE 2016; Tamura et al. 2019). According to tide gauge readings, mean sea level increased 2.45 mm/year between 1960 and 2014; satellite data indicates a 3.50 ± 0.7 mm/year rise between 1993 and 2014 (MONRE 2016). As sea levels continue to rise, those living in and near coastal areas face a serious threat.

Figure C3.1 shows the potentially inundated areas of Vietnam due to SLR in 2100 under the RCP8.5 scenario, as well as the distribution of mangrove forests. The Mekong River Delta (MRD) and Red River Delta were projected to experience considerable inundation because of their low elevations. The total area of the MRD is about 40,000 km^2 and was inhabited by nearly 18 million people in 2018. The elevation of coastal areas in the MRD ranges from 0.3–0.7 meters and the length of coastal area is more than 1,300 km. In Vietnam, sea dikes, mainly made of soil, were commonly constructed in the 1980s. Concrete structures of several types have also been put in place in coastal zones. However, the construction techniques used for these structures have tended to be simple and primitive and the barriers are susceptible to collapse (Tamura et al. 2018). Figure C3.2 shows the changes in the coastline at the town of Vinh Chau in Soc Trang Province based on images obtained by Google Earth in January 2006 compared to those taken by a unmanned aerial vehicle (UAV) in November 2015. Many of the mangrove forests have disappeared and

M. Tamura (✉)
Global and Local Environment Co-creation Institute, Ibaraki University, Ibaraki, Japan

VNU Vietnam Japan University, Hanoi, Vietnam
e-mail: makoto.tamura.rks@vc.ibaraki.ac.jp

O. T. Pham
VNU Vietnam Japan University, Hanoi, Vietnam

© The Author(s) 2022
T. Ito et al. (eds.), *Interlocal Adaptations to Climate Change in East and Southeast Asia*, SpringerBriefs in Climate Studies,
https://doi.org/10.1007/978-3-030-81207-2_13

Fig. C3.1 Inundated area
due to SLR by 2100 under
the RCP8.5 (MIROC-
ESM) scenario and
mangrove distribution

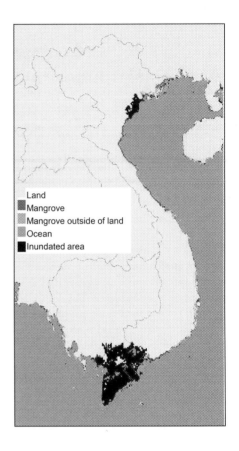

areas of the coastline showed extensive erosion at some locations due to a decrease
in sediments from upstream river reaches, economic activities, and long-term SLR.

Adaptation measures can reduce the adverse impacts of climate change. There
are three main approaches to adapt to the impact of SLR in coastal areas: retreat,
accommodation, and protection. For reliable adaptations in the future, it is necessary to examine the effectiveness and cost of coastal protection. This study evaluated the effectiveness of adaptation scenarios in the MRD—one involving dikes
only, the other combining dikes and mangrove planting—and estimated their costs.
The coastline of the MRD is 1,302 km long and includes 945 km of mangrove forest
in estuary areas.

The total population at risk of SLR in the MRD may increase roughly early in
this century and then decrease during the rest of twenty-first century under five
socio-economic scenarios (Fig. C3.3). Figure C3.4 shows an overview of economic
losses due to SLR in the MRD without adaptation under five Shared Socioeconomic

Fig. C3.2   Change in coastlines at the commune of Vinh Chau, Soc Trang (Tamura et al. 2018)

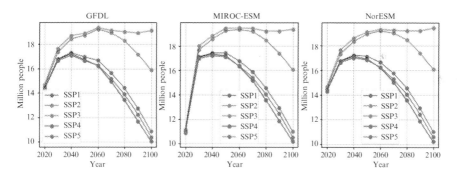

Fig. C3.3   Total population at risk in the MRD in the twenty-first century (Pham et al. 2020)

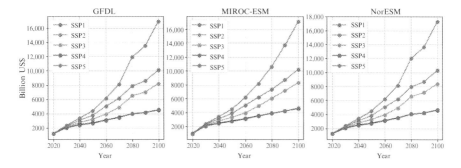

Fig. C3.4   Economic damage in MRD in the twenty-first century (Pham et al. 2020)

Table C3.1 Comparison of mixing grey and green infrastructures with other adaptations (discount rate = 3%) Unit: billion US$

		SSP1	SSP2	SSP3	SSP4	SSP5
Mixing grey and green	Benefit	3,303	2,809	2,133	2,112	4,405
	Cost	16	14	12	12	19
Earth dike and mangrove	Benefit	3,308	2,813	2,136	2,115	4,411
	Cost	22	19	16	16	26
Concrete dike	Benefit	3,290	2,798	2,124	2,103	4,389
	Cost	16	14	13	12	18

Pathways (SSP1-5). The economic damage was estimated by multiplying the inundated area and the value of land.

The study also presents a cost benefit analysis derived from mixing grey and green infrastructures in the MRD to adapt to SLR. It was assumed that the area's existing earth dikes will be upgraded to a height of 4 m and no mangrove and concrete dikes will be built in the area. Mixing grey and green infrastructures is evaluated for the following four scenarios: (1) A concrete dike system with a height of 4 m is built in areas without mangroves; (2) Mangrove forests can grow in suitable areas; (3) The lifespan of concrete dikes is 100 years; (4) The lifetime of mangrove is 50 years and the growth period of mangrove forests is 10 years before they can perform a protection function.

Table C3.1 compares the effectiveness of mixing grey and green infrastructures with other adaptation options. The benefit of earth dikes in combination with mangrove forests, or mixing grey and green infrastructures, is higher than the benefit from concrete dikes alone. The results also indicate that the benefits of SLR adaptation can exceed the costs required to set up an adaptation system. It is found that combining mangrove forests and dikes can serve as an effective climate change adaptation measure in Vietnam, where the potential inundated area and economic damage may be quite large.

References

MONRE (2016) Climate change and sea level rise scenarios for Viet Nam. Ministry of Natural Resources and Environment

Pham TO, Tamura M, Kumano N, Nguyen Q (2020) Cost-benefit analysis of mixing grey and green infrastructures to adapt to sea level rise in the Vietnam Mekong River Delta. Sustainability 12(24):19

Tamura M, Yasuhara K, Ajima K, Trinh VC, Pham SV (2018) Vulnerability to climate change and Residents' adaptations in coastal areas of Soc Trang Province, Vietnam. Int J Glob Warm 16(1):102–117

Tamura M, Yotsukuri M, Kumano N, Yokoki H (2019) Global assessment of the effectiveness of adaptation in coastal areas based on RCP/SSP scenarios. Clim Chang 152(3–4):363–377

Column 4
The Effect of Climate Change and Natural Disasters on Mangrove Forests in Xuan Thuy National Park: Proposed Adaptation Measures for Mangrove Forests

Quang Van Nguyen

Xuan Thuy National Park (XTNP), the first Ramsar site in Southeast Asia, is located in a low-lying coastal region in Nam Dinh Province of Vietnam. Most of the core zone of the XTNP is comprised of mangrove forests, which are inhabited by 222 indigenous and migratory bird species, 202 plant species, 386 invertebrate species, and 154 fishes (Fig. C4.1). Many of these species are rare, including nine endangered birds that are listed in the IUCN Red List of Threatened Species, have been recorded and sighted in the XTNP. The endangered birds are Western Curlew, Black-faced Spoonbill, Saunders's Gull, Painted Stork, Asian Dowitcher, Spoon-billed Sandpiper, Spot-billed Pelican, Nordmann's Greenshank, and Chinese Egret (MONRE 2020).

The XTNP supports several key mangrove plants species, such as *Aegiceras corniculata*, *Sonneratia caseolaris*, and *Kandelia obovata* (MONRE 2020). Mangrove forests in the XTNP also help to protect the sea dykes and provide livelihoods for thousands of people in the buffer zone of the park. However, due to the impacts of climate change, sea level rise, and other natural disasters, mangrove forests in many areas of the XTNP have been degraded.

Lecturers and students in the master's program in Climate Change and Development (MCCD) at VNU Vietnam Japan University conducted fieldwork and surveys in the XTNP in October 2020 (MCCD 2020). The results of the survey indicated that many mangroves were damaged or killed by coastal erosion and waves along a 7.5 km coastal stretch of the XTNP (Fig. C4.2). The direct impact of the erosion and strong waves reduced the amount of soil under the trees; the trees' roots lost their grip and the stumps moved. In addition, the high tides also brought sand from the ocean into the mangrove forests, which buried the roots of mangrove trees for extended periods and causing the death of the mangroves.

Q. Van Nguyen (✉)
VNU Vietnam Japan University, Hanoi, Vietnam
e-mail: quangmda@gmail.com

© The Author(s) 2022
T. Ito et al. (eds.), *Interlocal Adaptations to Climate Change in East and Southeast Asia*, SpringerBriefs in Climate Studies,
https://doi.org/10.1007/978-3-030-81207-2_14

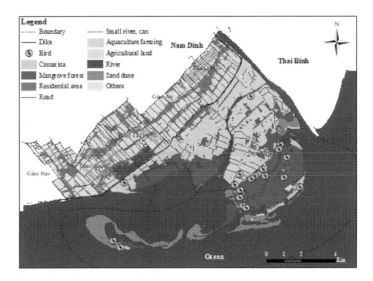

Fig. C4.1 Mangrove forests in the XTNP

Fig. C4.2 Dead mangroves caused by coastal erosion and waves

The loss of mangrove forests in the XTNP may be linked to climate change and the increased incidence of natural disasters. Interviews conducted with national park leaders, local government leaders, local residents, and statistical data show that the number and intensity of typhoons and extremely hot days has increased in recent years in the XTNP (MCCD 2020). The natural disasters and extreme weather have led to branch breakage and tree death, and indirectly reduced resistance to pests and diseases, resulting in widespread degradation and the loss of natural recovery capacity (Tran et al. 2019).

As the XTNP is located in a low-lying coastal region, the park could be at risk from sea level rise. Research conducted by Nguyen (2019) indicated that mangrove forests in the XTNP could be inundated by 1 m or 2 m of sea water due to sea level rise, and the affected mangrove forests could disappear in the future. As shown in Fig. C4.3, if the sea level rises 1 m or 2 m, the area of mangrove forests could be reduced by 21% or 52.2%, respectively (Nguyen 2019).

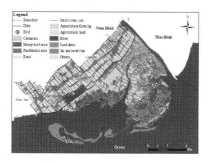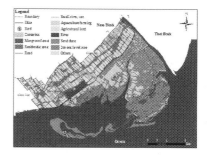

Fig. C4.3 Impacts of sea level rise scenarios on mangrove forests in the XTNP (left: 1 m; right: 2 m sea level rise)

In conclusion, mangrove ecosystems in the XTNP are threatened by climate change and natural disasters. Large areas of mangrove forests may disappear in the future due to sea level rise. Some adaptation solutions for protecting mangrove forests in the XTNP include: (a) casuarina trees should be planted extensively on sand dunes outside of mangrove forests to protect mangroves and other species from coastal erosion, storms, waves, and sea level rise; (b) mangrove species that are more resistant to the effects of extreme weather and natural disasters should be planted to replace degraded mangrove forests, or planted on bare land; and (c) long-term strategies and plans prepared by the local government and leaders of the XTNP in relocating and expanding mangrove forests to higher-elevation areas in the north and west of the park in order to mitigate against future sea level rise and other disasters.

References

MCCD (2020) A fieldwork and survey conducted by MCCD in XTNP in October 2020. Vietnam Japan University

MONRE (2020) Current status of biodiversity of Xuan Thuy National Park. The Ministry of Natural Resources and Environment, 199p (in Vietnamese)

Nguyen Q (2019) Development of a geospatial model to assess the impact of sea level rise on Xuan Thuy National Park. The project was funded by Vietnam Japan University. Vietnam Japan University, 45p (In Vietnamese)

Tran TMS, Nguyen TKC, Le HL, Tran VH, Pham TQ, Nguyen TTV, Pham TD (2019) Current status of mangroves in the context of Climate Change in Xuan Thuy National Park Buffer Zone, Nam Dinh Province. In: Vietnam international conference on Asian and Pacific coasts, pp 1221–1228

..

Column 5
Climate Change and Crop Management in Indonesia

Eri Matsuura and Nobuo Sakagami

Rice is a daily staple for the most of Indonesian people, and rice consumption has increased every year as the population has increased (Suryani et al. 2016). Climate change has severely affected rice production by increasing minimum temperatures and changing rainfall patterns. The decline in rice yield due climate change could threaten national food security over the long term. To maintain the food self-supporting rate and support farmers, new technologies for adapting to climate change in farming systems are needed. Asia-Pacific Climate Change Adaptation Information Platform (AP-PLAT) (2020) reported that the Indonesian government published the National Action Plan for Climate Change Adaptation (*Rencana Aksi Nasional Adaptasi Perubahan Iklim* or RAN-API) in 2014, and the Ministry of the Environment in Japan (MOEJ) and the Ministry of National Development Planning, Indonesia (BAPPENAS) are cooperating to assess the impact of climate change for local adaptation planning in the Republic of Indonesia.

Several farming systems have been proposed for adapting to climate change in Indonesia. First, the System of Rice Intensification (SRI) was developed in the early 1980s by Fr. Henri de Laulanie (SRI-Rice 2020). SRI is based on four main principles: (1) early plant establishment; (2) reduced plant density; (3) improved soil conditions through enrichment with organic matter; and (4) reduced and controlled water application (SRI-Rice 2020). To improve rice production, SRI was introduced in 1999 in Indonesia. The SRI method not only increases rice yields by using less water, grain and fertilizer, but it can also contribute to greenhouse gas mitigation from paddies.

In addition, according to a report based on knowledge exchange on the rice farming practices in Yogyakarta, Indonesia (FAO 2016), conservation farming systems

E. Matsuura · N. Sakagami (✉)
College of Agriculture, Ibaraki University, Ibaraki, Japan
e-mail: nobuo.sakagami.soil@vc.ibaraki.ac.jp

© The Author(s) 2022
T. Ito et al. (eds.), *Interlocal Adaptations to Climate Change in East and Southeast Asia*, SpringerBriefs in Climate Studies,
https://doi.org/10.1007/978-3-030-81207-2_15

such as the *Mina Padi* system (Fig. C5.1) and the *Jajar Legowo* planting system (Fig. C5.2) have been practiced by local farmers.

Balai Pengkajian Teknologi Pertanian (BPTP) Sulawesi Barat (2020) described the *Mina padi* system as an integrated farming system for cultivating fish in paddies. Farmers derive four benefits from this system: (1) Minimum use of fertilizer; (2) Fish are natural predators of the pests in paddies; (3) Save time and labor due as weeding performed by fish; and (4) Double income by selling fish. Another integrated farming system, *Jajar Legowo* planting system has some variations that are used to cultivate rice using different spacing of transplanted rice, such as *Jarwo* 2:1, 3:1 or 4:1. Ducks are used for pest and weed control. *Legowo* 2:1 method can increase rice productivity by 1.5 times compared to the SRI method (Darmawan 2016). Regular spacing can create suitable humidity and temperature for rice growth in paddies.

Fig. C5.1 *Mina padi* system (BPTP Sulawesi Barat 2020)

Fig. C5.2 *Jajar Legowo* planting system 4:1 (Fausayana and Tarappa 2018)

Kusumasari (2016) noticed that farmers understand climate change occurring in their region and its influences on their cultivation method. She also pointed out that changing the planting pattern, using soil cultivation technique, plant pest management technique, and watering/irrigation technique can be used as countermeasures. Various integrated farming systems might be resilient to climate change and sustainable by saving water, less use of fertilizers and increasing soil organic matter. Moreover, farmers can manage the system better and adapt to the local environment in different islands of Indonesia. Farmers know the best farming practices in their region based on experience. Climate-smart agricultural technologies do not need to be expensive; rather, farmers' experience and new knowledge can be combined to produce better outcomes in farmers' livelihoods and mitigate global warming.

References

AP-PLAT (2020) Assessing the impact of climate change for local adaptation planning in the Republic of Indonesia. https://ap-plat.nies.go.jp/plan_implementation/international/Indonesia/index.html

BPTP Sulawesi Barat (2020) Teknologi Mina Padi Dengan Sistim Tanam Jajar Legowo. http://sulbar.litbang.pertanian.go.id/ind/index.php/info-teknologi/344-teknologi-mina-padi-dengan-sistim-tanam-jajar-legowo (In Indonesian)

Darmawan M (2016) Analysis of *Legowo row* planting system and system of rice intensification of paddy field (*Oryza sativa* L.) toward growth and production. AgroTech J 1:14–18

FAO (2016) Report of the knowledge exchange on the promotion of efficient rice farming practices, farmer field school curriculum development and value chains. FAO Fisheries and Aquaculture Report No 1181. http://www.fao.org/3/a-i6617e.pdf

Fausayana I, Tarappa W (2018) The technology adoption of jajar legowo system and direct seeding system on rice farming in the village of Duria Asi, Wonggeduku District of Konawe regency, Indonesia. Int J Bus Manag Technol 2:28–33

Kusumasari B (2016) Climate change and agricultural adaptation in Indonesia. MIMBAR J Soc Dev 32:243–353

SRI-Rice (2020) SRI International Network and Resources Center. http://sri.ciifad.cornell.edu/

Suryani N, Abdurrachim R, Alindah N (2016) Analysis of carbohydrate, fiber and glycemic index of processed rice cultivar *Siam Unus* as an alternative snack for diabetes mellitus. Indonesian J Health 7:1–9. (In Indonesian)

Column 6
Stakeholder Perception and Empirical Evidence: Oil Palm Biomass Utilization as Climate-Smart Smallholder Practice

Sachnaz Desta Oktarina, Ratnawati Nurkhoiry, Rizki Amalia, and Zulfi Prima Sani Nasution

Little is known about smallholder perception and adaptation of climate-smart practices. More precisely, smallholder's oil palm plots were frequently accused of being the driving force of ecosystem service depletion. This was arguably due to the vicious cycle facing the smallholder as the push factor. At first, smallholders did not plant legitim seeds and were not equipped with good agricultural practices (GAP), which decreased productivity. Since their earnings are low, they are motivated to expand the farm through deforestation (Bennett et al. 2018). This low-income situation is mainly due the absence of other income-generating activities to support their livelihoods. In the long run, this kind of practice has caused climate change. Scientists have suggested that climate-smart agricultural (CSA) practices are applied by smallholders in order to overcome this problem. Fortunately, palm oil plantation has offered a pull factor to answer this challenge. Palm oil biomass utilization might act as a pull factor that can be performed to the smallholder context. Specifically, palm oil products are cheap and eco-friendly because they have ubiquitous and versatile functions, ranging from food to fuel. In addition, little is known that we still can make the most of its left-over as well. Indonesian oil palm smallholders might apply oil palm biomass to grow food; make crafts; or use as wood, fiber, and energy. Thus, the versatility of palm oil definitely mitigates depletion of the environment and its use can be expanded to other smallholders nationwide. The act of using biomass is also considered to be a CSA practice that is specific to oil palm plantations. While most studies have examined CSA in terms of mechanistic or biophysical aspects, relatively little research has examined the interest, perception, or acceptance of CSA in oil palm cultivation, particularly among smallholders.

This research was supported by INSINAS Research Grant No: 18/INS-1/PPK/E4/2019.

S. D. Oktarina (✉) · R. Nurkhoiry · R. Amalia · Z. P. S. Nasution
Indonesian Oil Palm Research Institute, Kota Medan, North Sumatera, Indonesia
e-mail: sachnazdo@iopri.org

© The Author(s) 2022
T. Ito et al. (eds.), *Interlocal Adaptations to Climate Change in East and Southeast Asia*, SpringerBriefs in Climate Studies,
https://doi.org/10.1007/978-3-030-81207-2_16

A study to assess stakeholder perception and the implementation of biomass utilization as a CSA by smallholders was performed in case studies in Indonesia. The Labuhan Batu (n = 24), Batu Bara (n = 29), Langkat (n = 14), and Serdang Bedagai (n = 32 smallholders) districts were selected as sample cases in North Sumatra Province where numerous oil palm concessions are located (Ministry of Agriculture Decree no. 833/KPTS/SR.020/M/12/2019). The first phase of the study was conducted by quasi-qualitative text mining and sentiment analysis (Pang and Lee 2008) to decode smallholder, practitioner, and expert's perceptions and sentiments regarding oil palm biomass products in small-medium enterprise (SME) schemes. The next phase of implementation was undertaken by introducing farmers to oil palm biomass-derived products, such as empty fruit bunch briquettes, oil palm frond pellets, midrib handicraft, oil palm based-livestock feed, empty fruit bunch oyster mushrooms, oil palm juice brown sugar, oil palm-laminated wood, and empty fruit bunch-compost. The feasibility and preferences among these alternatives were then assessed by multi-criteria decision-making tools in what is referred to as an Analytical Hierarchy Process (AHP) based on the benefit (B), opportunity (O), cost (C), and risk (R) features, as illustrated in Table C6.1. The text mining analysis revealed that, initially, the smallholders were perceived to have a lower interest in making use of oil palm biomass. It was suggested that they expected a higher incentive to adopt it in smallholder groups (Kehinde et al. 2019). It was also still unclear whether they realize and understand the potential of biomass utilization to save the environment and restore ecological services. The word cloud shown in Fig. C6.1 showed that none of the smallholders considered either sustainability or environmental services. They also had little enthusiasm towards the project since they depend on the mini palm oil mill more than the biomass utilization itself. They were reluctant to employ only biomass utilization as it perpetuates smallholder marginalization. Moreover, aside from their concerns related to accessing feedstock and raw materials, they also anticipated obtaining income immediately after the biomass utilization practices were conducted. However, the other two stakeholders revealed

Table C6.1 Biomass utilization alternatives in view of its benefit, opportunity, cost, and risk

Criteria	Sub-criteria	Briquette	Handicraft	Mushroom	Oil Palm Sugars	Compost
B	Income	Rp 18 M/ month	Rp 5 M/ month;	Rp 9 M/ month;	Rp 23.5 M/ month;	Rp 20 M/ month;
	Labour	3 people	5 people	2 people	4 people	> 10 people
O	Raw material	5 MT/ha/ year	350 kg/ha/ year	5 MT/ha/ year	7–9 L/tree	5 MT/ha/ year
	Market	Domestic use fuel	Creative industry	Healthy food	Oil palm sugars	Organic fertilizer
C	Investment	Rp 70 M	Rp 5 M	Rp 30 M	Rp 35 M	Rp 600 M
	Operational	Rp 18 M/ month	Rp 7.4 M/ month	Rp 4.5 M/ month	Rp 25 M/ month	Rp 40 M/ month
R	Market	High	High	Low	Low	Low
	Environment	Low	Low	Low	Low	Low

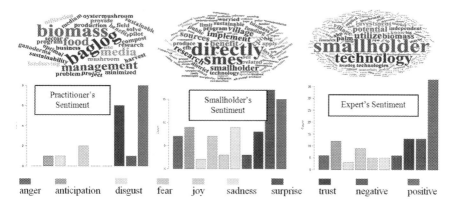

Fig. C6.1 Word cloud and sentiment analysis results of stakeholder perception towards biomass utilization

that they had generally positive sentiments towards the CSA of biomass utilization. The findings showed that the two groups were enthusiastic and showed considerable awareness towards the projects. The practitioners shared their experiences in how it could be successfully executed while maintaining nature's carrying capacity. The experts were convinced of the benefits and feasibility of the program to be applied to SME that is suitable with smallholder's institutional arrangement. They also highlighted the importance of low investment cost, the site-specific potential, and the need for alternative income-generating activities as these related to biomass utilization. Overall, both practitioners and experts agreed to provide assistance to smallholders and facilitate access to raw materials and a capacity building program for them. Eventually, immediately after implementation, the smallholders were enlightened and selected biomass utilization technology as a pluri-activity. The AHP results (ordinal ranking) suggested that midrib handicrafts were dominant, followed by the oyster mushroom, empty fruit bunch compost, briquettes, and oil palm sugars were most feasible products according to the smallholder's point of view (AHP scores: 0.228, 0.139, 0.128, 0.122, and 0.117 respectively). It showed the midrib handicraft was mostly selected due to its lower investment costs, despite returning the lowest disposable incomes (5 million IDR per month) compared to other biomass products. Since these shareholders do not generate enough money to invest, they tend to be risk-averse in their selection. However, it is necessary to provide an incentive scheme that might encourage smallholders to adapt. In doing so, the climate-smart adaptation practices are considered feasible and expandable and can be applied nationwide for the sustainability of the people, the profit, and the planet.

References

Bennett A, Ravikumar A, Paltán H (2018) The political ecology of oil palm company-community partnerships in the Peruvian Amazon: deforestation consequences of the privatization of rural development. World Dev 109:29–41

Kehinde MO, Shittu AM, Osunsina IOO (2019) Willingness to accept incentives for a shift to climate-smart agriculture among lowland Rice farmers in Nigeria. Nigerian J Agric Econ 9(2066-2020-1439):29–44

Pang B, Lee L (2008) Opinion mining and sentiment analysis. Found Trends Inf Retr 2(1–2):1–135

Column 7
Water Resource Assessment and Management in Phuket, Thailand

Sukanya Vongtanaboon

Within the context of climate change, the province of Phuket in Thailand is the only province in Thailand without any large rivers, which means that it is potentially at risk from problems associated with water scarcity in the future. Phuket Island is also the largest island in Thailand with an area of 543 km^2 comprised of three districts (Thalang, Muang and Kathu); 77% of the island is mountainous and 23% consists of plains, primarily in the central and eastern regions. Most water is derived from rainfall, which is stored in reservoirs, mine shafts, and groundwater sources. Due to geography, Phuket is a popular tourist destination. Rates of development in the hotel industry and residential areas, as well as the population, have been rapid in recent decades. However, these developments have severely degraded watershed areas. Economic growth and increases in the number of tourists has resulted in an increase in demand for water, and in the year 2020 consumption exceeded 100,000 m^3/day (Provincial Waterworks Authority 2020). Moreover, trends of water use in Phuket are expected to increase by 12% annually, and it is forecasted that the water demand for local consumption and tourism will rise to 103.07 million m^3/year in 2032 (Patong Municipality 2020). As the water supply in the three major reservoirs (Bang Wad, Bang Niew Dam and Klong Katha) are limited due to long periods of a dry weather (Provincial Waterworks Authority 2020), there are problems of water scarcity in Phuket and other areas with public water systems in the province. The droughts in 2002 and 2019–2020 caused severe water scarcity in Phuket, and this has become an important issue that the province needs to address.

Rainfall Phuket typically experiences a tropical monsoon climate with a seven-month dry season (mid-October to mid-May) with occasional rain, and a five-month rainy season. A study of monthly rainfall in Phuket found that the rainfall was low-

S. Vongtanaboon (✉)
Faculty of Science and Technology, Phuket Rajabhat University, Phuket, Thailand
e-mail: sukanya.v@pkru.ac.th

© The Author(s) 2022
T. Ito et al. (eds.), *Interlocal Adaptations to Climate Change in East and Southeast Asia*, SpringerBriefs in Climate Studies,
https://doi.org/10.1007/978-3-030-81207-2_17

Fig. C7.1 Monthly rainfall (mm) of Thalang, Muang, Kathu districts and average monthly rainfall in Phuket

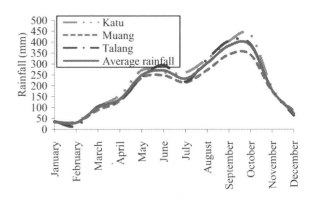

est in January–February (30–35 mm) and highest in August–October (303–385 mm). Monthly rainfall increases from March to October. Figure C7.1 shows monthly rainfall of the three districts and average monthly rainfall of Phuket. Annual rainfall of Phuket is in the range of 1882.7-2713.5 mm, with the average value of 2200 mm/year. The annual rainfall was below average in 1997, 2002, 2005, 2019–2020, which might be caused by the El Nino phenomenon (Golden Gate Weather Services 2020).

Groundwater Groundwater resources in Phuket can be classified into two types based on the degree of compaction and cementation, which are unconsolidated aquifers and consolidated aquifers (Department of Mineral Resources 2020). An evaluation of groundwater resources showed that 5.11 million m^3 was extracted from 1024 groundwater wells.

Surface Water Surface water in Phuket consists of reservoirs, mine shafts, peat bogs, and canals (Irrigation Office 2020). The major source of water supply in Phuket is from the Bang Wad Watershed located in Kathu District (Fig. C7.2). The main canals are the Bangyai, Krang and Tha Yang canals, which have annual discharges of 49.02, 45.17 and 45.09 million m^3, respectively. The average annual discharge is 153.29 million m^3. Water storage capacity in the three main reservoirs of Phuket, i.e., Bang Wad, Bang Niew Dam and Klong Katha were 10.20, 7.20 and 4.32 million m^3 respectively. In addition, the water storage capacity of 109 old tin mines was 21.02 million m^3.

Water Resource Management The problem of water shortage due to climate change and tourism development of Phuket requires good planning and management. The main issues related to water resource management in Phuket are as follows: (1) insufficient water supply for consumption by the local population and increasing inbound tourists; (2) incomplete area coverage by Phuket Provincial Water Works Authority; (3) lack of land to expand the main water lines; and (4) lack of budget for water resource development and service area expansion. To resolve the water supply problem, it is necessary to identify potential water sources, including feasibility studies in various fields that will resolve the problem of water shortages

Fig. C7.2 Bang Wad Watershed, Phuket

during the dry season. In conclusion, the solutions for mitigating water shortages in Phuket are as follows: (1) increase water storage: development of small irrigation projects, old tin mines and peat swamps; (2) improve the water retention capacity of main reservoirs; (3) provide more water to main reservoirs; (4) develop a water drainage system between main reservoirs; (5) develop groundwater pumping; and (6) transfer water from outside of Phuket (Phuket Provincial Administration Organization et al. 2020).

References

Department of Mineral Resources (2020) Groundwater map guide book: Phuket Province. Groundwater Division, Bangkok

Golden Gate Weather Services (2020) El Niño and La Niña years and intensities. https://ggweather.com/enso/oni.htm

Irrigation Office (2020) Surface water in Phuket. Phuket Irrigation Project, Phuket

Patong Municipality (2020) Phuket water and wastewater plans, Phuket

Phuket Provincial Administration Organization, Royal Irrigation Department, Provincial Waterworks Authority, Phuket City Municipality, Ministry of Natural Resources and Environment, Phuket Rajabhat University, Prince of Songkla University (2020) Water management in Phuket

Provincial Waterworks Authority (2020) Water source. Water resource conservation, Phuket

Column 8
Analysis of Measures for Preventing Desertification in Inner Mongolia in China

Yulu Ma

Since the early 1990s, with the economic development and population growth, desertification has increasingly attracted the attention of governments, international organizations, and scientists around the world. The evaluation of desertification has become a new interest in land science research. Inner Mongolia is a largely deserted province in China, with the deserted land covering 640,000 km^2; 91.16%, 41.4%, and 41.0% of the desertification was caused by wind erosion, water erosion, and saline desertification, respectively. Desertification is so disastrous that it can cause additional ecological problems (Liu and Wang 2006; Ye 2008). The causes of desertification in Inner Mongolia can be classified into human and natural causes. Regarding human causes, irrational irrigation methods are the main causes, mainly due to the pressure of population growth, overgrazing (Fig. C8.1), the expansion of dry land reclamation (Fig. C8.2), woodcutting, and the harvesting of Chinese herbal medicine (Fig. C8.3). Regarding natural causes, climate change and the geographical environment are the main issues. Within Inner Mongolia, the arid, semi-arid, and sub-humid arid regions deep in the hinterland of the continent and far away from the ocean comprise the most arid and fragile environmental zone, which lies in the same latitude as areas with the lowest precipitation and highest evaporation. In the past 40 years, precipitation has shown a decreasing trend in parts of the arid, semi-arid, and sub-humid arid regions of Inner Mongolia, whereas the temperature in other areas has shown an increasing trend. These changes in the climate have led to an increase in evaporative power and contributed to soil salinization, which have exacerbated desertification to a certain extent.

To avoid desertification, several measures can be taken, including planting new vegetation, practicing reasonable water use, and strengthening ecological agriculture construction. The favorable conditions of light and heat in the arid and

Y. Ma (✉)
College of Agriculture, Inner Mongolia Minzu University, Tongliao, China
e-mail: yuluma@imun.edu.cn

© The Author(s) 2022
T. Ito et al. (eds.), *Interlocal Adaptations to Climate Change in East and Southeast Asia*, SpringerBriefs in Climate Studies,
https://doi.org/10.1007/978-3-030-81207-2_18

Fig. C8.1 Overgrazing

Fig. C8.2 Over-reclamation

semi-arid areas provide sufficient light for the growth of forests. However, the
healthy growth of forests is greatly hindered because of droughts, water shortages,
and the dry climate. Therefore, technical measures of sand control and afforestation,
seedling cultivation, water-saving, and drought resistance should be developed to
create a favorable environment for the growth of forests. Energy should be devoted
to evoke the positive impacts of such measures and promote the healthy growth of

Fig. C8.3 Harvesting of Chinese herbal medicine

forests. Therefore, the advancement of technologies for vegetation and afforestation should be promoted in arid and semi-arid areas to improve the growth of forests (Bao 2002). Water is the key factor in the prevention of desertification. However, the prevention of desertification will be hindered because of the lack of adequate water. We should therefore improve farming and irrigation techniques and advocate for water-saving agriculture to avoid soil salinization in farming areas. The number of wells should be reduced to avoid the uncontrolled growth of livestock on pastoral grasslands. As for the arid inland, water resources, including the water upstream and downstream the river should be rationally allocated, and water-saving projects should be carried out. The cultivated land subjected to heavy wind erosion and desertification, consisting of low levels of organic matter and limited groundwater resources and characterized by unsuitable farming conditions, should be converted into grassland, where the restoration of vegetation should be sped up through a combination of engineering and natural restoration to recover the ecological functions of this area.

The increasing population is also one of the main factors explaining the desertification of this area because the demand for food and firewood has also increased accordingly. Therefore, the large degree of reclamation and the expansion of arable land have caused the increasing destruction of land resources and accelerated the development of desertification. The lesson in regard to exceeding the supportive capacity of the land for the population is profound. Evidence has shown that humans cannot demand from nature without limits. Based on this understanding, controlling the population and implementing family planning have become part of basic national policy (Yu 2003; Ye 2008).

References

Bao Q (2002) Status quo analysis and countermeasure research on desertification in Inner Mongolia. Inner Mongolia Soc Sci Chinese Edition 6:56–60. (in Chinese)

Liu A, Wang G (2006) Analysis of the causes of land desertification in Inner Mongolia [A]. Proceedings of 2006 China Grass Industry Development Forum, 2006-9-1, pp 332–339 (in Chinese)

Ye R (2008) The status quo of land desertification in Inner Mongolia and its control countermeasures. Western Resour 6:37–39. (in Chinese)

Yu Y (2003) Analysis of Inner Mongolia desertification policy factors since the Republic of China. J Inner Mongol Normal Univ (Philosophy and Social Science Edition) 3:79–83. (in Chinese)

Column 9
The Power of Dialogical Tools in Participatory Learning

Akihiko Kotera

How Can We Promote Dialogue?

It seems that most of the complex problems in the world are made difficult to solve by a lack of dialogue. Needless to say, "dialogue" does not immediately solve everything, but at least we can find much more hope compared with a situation without dialogue.

Participatory learning is one method for acquiring new knowledge and skills (Fig. C9.1). It is a method that emphasizes the importance of encouraging participants to take spontaneous actions such as experiencing and interacting and to learn from the process by themselves. In reality, however, it is quite difficult to establish a rich dialogue between participants and the facilitators of participatory learning.

One of the factors that inhibit dialogue in participatory learning is the means of communication of scientific knowledge and information. Various devices and tools, such as geographic information maps, illustrations, and models, have been used to convey information in an easy-to-understand manner. However, in many cases, the very tools that are supposed to provide useful information unintentionally interfere with the dialogue with the participants because their expertise is not translated well. This is because the act of reading and understanding geographical maps and graphs correctly is a specialized skill that requires a certain amount of training. In addition, most tools are designed for one-way transmission of information and have not been designed to be used in dialogue. In this context, new information communication tools/interfaces such as "Tangible Bit" are expected to be utilized.

A. Kotera (✉)
Global and Local Environment Co-creation Institute, Ibaraki University, Ibaraki, Japan
e-mail: akihiko.kotera.moctuyen@vc.ibaraki.ac.jp

© The Author(s) 2022
T. Ito et al. (eds.), *Interlocal Adaptations to Climate Change in East and Southeast Asia*, SpringerBriefs in Climate Studies,
https://doi.org/10.1007/978-3-030-81207-2_19

Fig. C9.1 Scene of
participatory learning

Fig. C9.1 Scene of
participatory learning

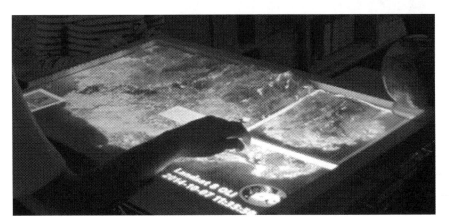

Fig. C9.2 Appearance of the tangible GIS

Designing the Tool

"Tangible Bit" consists of a 1:7500 scale three-dimensional (3D) terrain screen, a
projector, a tablet PC, and an infrared camera-based touch sensor, and can project
various information on the 3D terrain of the target area (Fig. C9.2). This system can
be assembled and moved easily in the field. The table-top style allows participants
to surround the 3D terrain screen naturally at a close distance from each other without creating physical barriers to access and interact with the information. The information contents to be displayed include satellite remote sensing images, various
GIS thematic maps such as land use maps, water management organization charts,

videos, and scanned images of existing paper maps, which are necessary and easy to learn. Each image has been geometrically corrected to fit the undulations of the 3D screen.

By superimposing these contents as needed and projecting them on the 3D topographic screen, unlike conventional flat maps, users can intuitively grasp various contents as well as topographic and hydrological information such as mountain slopes, height differences, and water flow directions. Furthermore, by touching and sliding your finger on the surface of the 3D screen, users can sense the information intuitively.

Trying It Out

The tool has been used in the field of collaborative water resources management with community participation in the Bilibili Irrigation command in southern Sulawesi and the Saba River Basin in northern Bali, Indonesia (Kubota 2016). The opportunities for using this tool include water user meetings, community stakeholder meetings, and meetings at universities and public institutions, and the information content used depends on the purpose of the meeting.

These meetings are held periodically by water users, where, for example, water allocation plans for irrigation in rice transplanting season are discussed. As a basis for discussing more rational water allocation, we analyzed the need to increase the common understanding of water resources in the entire region. The goal of the study was to understand (1) where and how much water used in the paddy fields is generated and how it flows, and (2) how the topography and environmental conditions of the paddy fields are distributed, both at the watershed scale and at the scale of each individual paddy field.

What Kind of Changes Did the New Tool Bring?

Although we have conducted similar learning using paper-based media in the past, the major change that we observed using this device was the atmosphere of the place, which clearly led to meaningful and prolonged "dialogue" (Fig. C9.3). Even members who do not usually speak up began to participate actively in the dialogue. This was triggered by the moment when each participant, including us, felt that they could intuitively understand the information (i.e., happy), and at the same time, were convinced that they were sharing the same feeling (i.e., fun).

Fig. C9.3 Scenes of participatory learning and mutual dialogue with the tangible GIS

What Have We Learned as Scientists?

It was also necessary for scientists to change our way of talking and our attitude toward dialogue to master the new tools. What we later realized was that the most significant change for us was also that participatory learning felt more fun than ever. The fun was in the serious dialogue with the participants, and we believe that the same feeling was conveyed and shared by the participants. At the same time, the tools provided information that was easy to understand, and also had a synergistic effect on the dialogue among all the participants, including us. This is the power of dialogical tools.

Reference

Kubota J (ed) (2016) Sharing the water resources. Bensei shuppan, Tokyo. (in Japanese)

Printed in the United States
by Baker & Taylor Publisher Services